Advances in Metal-Containing Magnetic Materials

Advances in Metal-Containing Magnetic Materials

Guest Editor
Zhongwu Liu

Basel • Beijing • Wuhan • Barcelona • Belgrade • Novi Sad • Cluj • Manchester

Guest Editor
Zhongwu Liu
School of Materials Science
and Engineering
South China University
of Technology
Guangzhou
China

Editorial Office
MDPI AG
Grosspeteranlage 5
4052 Basel, Switzerland

This is a reprint of the Special Issue, published open access by the journal *Metals* (ISSN 2075-4701), freely accessible at: www.mdpi.com/journal/metals/special_issues/metal_magnetic-materials.

For citation purposes, cite each article independently as indicated on the article page online and using the guide below:

Lastname, A.A.; Lastname, B.B. Article Title. *Journal Name* **Year**, *Volume Number*, Page Range.

ISBN 978-3-7258-3352-8 (Hbk)
ISBN 978-3-7258-3351-1 (PDF)
https://doi.org/10.3390/books978-3-7258-3351-1

© 2025 by the authors. Articles in this book are Open Access and distributed under the Creative Commons Attribution (CC BY) license. The book as a whole is distributed by MDPI under the terms and conditions of the Creative Commons Attribution-NonCommercial-NoDerivs (CC BY-NC-ND) license (https://creativecommons.org/licenses/by-nc-nd/4.0/).

Contents

About the Editor vii

Zhongwu Liu
Advances in Metal-Containing Magnetic Materials and Magnetic Technologies
Reprinted from: *Metals* **2023**, *13*, 1318, https://doi.org/10.3390/met13071318 1

Jiayi He, Jiali Cao, Zhigao Yu, Wenyue Song, Hongya Yu and Mozaffar Hussain et al.
Grain Boundary Diffusion Sources and Their Coating Methods for Nd-Fe-B Permanent Magnets
Reprinted from: *Metals* **2021**, *11*, 1434, https://doi.org/10.3390/met11091434 5

Caihai Xiao, Weiwei Zeng, Yongli Tang, Cifu Lu, Renheng Tang and Zhigang Zheng et al.
Effect of Ho Substitution on Magnetic Properties and Microstructure of Nanocrystalline
Nd-Pr-Fe-B Alloys
Reprinted from: *Metals* **2022**, *12*, 1922, https://doi.org/10.3390/met12111922 21

Qintian Xie, Hongya Yu, Han Yuan, Guangze Han, Xi Chen and Zhongwu Liu
Enhanced Magnetic Properties and Thermal Conductivity of FeSiCr Soft Magnetic Composite
with Al_2O_3 Insulation Layer Prepared by Sol-Gel Process
Reprinted from: *Metals* **2023**, *13*, 813, https://doi.org/10.3390/met13040813 30

Hongya Yu, Jiaming Li, Jingzhou Li, Xi Chen, Guangze Han and Jianmin Yang et al.
Enhancing the Properties of FeSiBCr Amorphous Soft Magnetic Composites by Annealing
Treatments
Reprinted from: *Metals* **2022**, *12*, 828, https://doi.org/10.3390/met12050828 44

Pan Luo, Hongya Yu, Ce Wang, Han Yuan, Zhongwu Liu and Yu Wang et al.
Properties Optimization of Soft Magnetic Composites Based on the Amorphous Powders with
Double Layer Inorganic Coating by Phosphating and Sodium Silicate Treatment
Reprinted from: *Metals* **2023**, *13*, 560, https://doi.org/10.3390/met13030560 52

Jiaming Li, Jianliang Zuo and Hongya Yu
Effects of La on Thermal Stability, Phase Formation and Magnetic Properties of
Fe–Co–Ni–Si–B–La High Entropy Alloys
Reprinted from: *Metals* **2021**, *11*, 1907, https://doi.org/10.3390/met11121907 66

Zongsheng He, Ziyu Li, Xiaona Jiang, Chuanjian Wu, Yu Liu and Xinglian Song et al.
Surface Investigation of $Ni_{81}Fe_{19}$ Thin Film: Using ARXPS for Thickness Estimation of
Oxidation Layers
Reprinted from: *Metals* **2021**, *11*, 2061, https://doi.org/10.3390/met11122061 74

Xi-Chun Zhong, Xu-Tao Dong, Jiao-Hong Huang, Cui-Lan Liu, Hu Zhang and You-Lin Huang et al.
One-Step Sintering Process for the Production of Magnetocaloric $La(Fe,Si)_{13}$-Based Composites
Reprinted from: *Metals* **2022**, *12*, 112, https://doi.org/10.3390/met12010112 83

Zijing Yang, Jiheng Li, Zhiguang Zhou, Jiaxin Gong, Xiaoqian Bao and Xuexu Gao
Recent Advances in Magnetostrictive Tb-Dy-Fe Alloys
Reprinted from: *Metals* **2022**, *12*, 341, https://doi.org/10.3390/met12020341 99

Chaoqun Xu, Li Yang, Kui Huang, Yang Gao, Shaohua Zhang and Yuting Gao et al.
Researching a Moving Target Detection Method Based on Magnetic Flux Induction Technology
Reprinted from: *Metals* **2021**, *11*, 1967, https://doi.org/10.3390/met11121967 126

Li Yang, Caihong Li, Song Zhang, Chaoqun Xu, Hun Chen and Shuting Xiao et al.
Correction Method of Three-Axis Magnetic Sensor Based on DA–LM
Reprinted from: *Metals* **2022**, *12*, 428, https://doi.org/10.3390/met12030428 . . . . **138**

About the Editor

Zhongwu Liu

Dr Zhongwu Liu is a full professor at the School of Materials Science and Engineering, South China University of Technology, China. He is currently the head of the Department of Metallic Materials Science and Engineering. He earned his Ph.D. degree from the University of Sheffield (UK). He was a post-doctorate fellow at the National University of Singapore and a research fellow at Nanyang Technological University (Singapore). His research interests are focused on magnetism, magnetic materials, and advanced metallic materials. In particular, he has been working on rare earth permanent magnets and soft magnetic materials and devices since the beginning of this century. He has published two books and about 500 research papers in peer-reviewed journals. His research has led to progress in the industries of RE-Fe-B hard magnets, amorphous/nanocrystalline magnetic alloys, and soft magnetic composites. He has received a number of awards for his excellent research and teaching.

Editorial

Advances in Metal-Containing Magnetic Materials and Magnetic Technologies

Zhongwu Liu

School of Materials Science and Engineering, South China University of Technology, Guangzhou 510640, China; zwliu@scut.edu.cn

1. Introduction

Magnetic materials generally refer to materials with ferromagnetic or ferrimagnetic ordering. In a broad sense, they also include weak magnetic and antiferromagnetic materials which can provide magnetism and a magnetic effect. Most magnetic materials contain metallic elements with 3d and/or 4f electrons, and they exhibit strong magnetism or significant interactions between magnetism and other physical properties.

Magnetic materials have found increasing applications in various fields, including electric motors, mechanical equipment, electronic devices, information recording, sensors, etc. The development of intelligent equipment, AI, 5G, consumer electronics, biomedicine, aerospace technology, and the military industry has created higher requirements for various types of magnetic materials. Since the previous century, traditional magnetic materials such as permanent magnets and soft magnets have seen innovations in their composition, structure, and processing. Advanced magnetic materials are continuously emerging. Concepts for new technologies are also under quick development.

This Special Issue entitled "Advances in Metal-Containing Magnetic Materials" has collected 11 manuscripts from a broad field of research. The topics addressed in this Special Issue include permanent magnets, soft magnets, magnetocaloric materials, magnetostrictive materials, magnetic thin films, and magnetic technologies. Nine research papers and two review papers provide the recent progress in magnetic materials and technology that has been made in the following institutes: South China University, the Guangdong Academy of Sciences, Southwest Minzu University, the University of Electronic Science and Technology of China, the University of Science and Technology Beijing, and the Beijing Institute of Spacecraft Environment Engineering. These articles are attractive for scholars in these fields.

2. Contributions

Two papers on permanent magnets are presented in this issue. He et al. [1] reviewed sources of grain boundary diffusion (GBD) and their coating methods for Nd-Fe-B permanent magnets. The GBD process, which is one of the most exciting technologies for rare earth permanent magnets that has emerged in this century, provides the best route for fabricating highly coercive Nd-Fe-B magnets with low levels of consumption of expensive HRE resources. Differing from previous review articles regarding GBD, this review provides an introduction of the typical types of diffusion sources and their fabrication approaches. The effects of the diffusion source on the microstructure and magnetic properties of magnets are summarized. In particular, the principles and applicability of different coating approaches ae discussed in detail. It is believed that this review can provide technical guidance for designing the diffusion process and products meeting specific requirements. Xiao et al. [2] report the effects of partially substituting the element Holmium (Ho) on the magnetic properties and microstructure of nanocrystalline melt-spun Nd-Fe-B alloys with a composition of $[(NdPr)_{1-x}Ho_x]_{14.3}Fe_{76.9}B_{5.9}M_{2.9}$ (M = Co, Cu, Al and Ga). They show that Ho can significantly enhance the coercivity (H_{cj}) and elevate the temperature behavior of the remanence (M_r) and the thermal stability. Their finding provides an important reference for the efficient improvement of the thermal stability of Nd-Fe-B-type materials.

Four papers focus on soft magnets. Li et al. [3] investigated the soft magnetic properties of the melt-spun, high-entropy alloys (HEAs) $Fe_{27}Co_{27}Ni_{27}Si_{10-x}B_9La_x$. They obtained superior soft magnetic properties with a low level of coercivity H_c of ~7.1 A/m and a high saturation magnetization B_s value of 1.07 T in an $Fe_{27}Co_{27}Ni_{27}Si_{9.4}B_9La_{0.6}$ alloy. They also found that the content of La has an important effect on the primary crystallization temperature and the secondary crystallization temperature of the alloys. Yu et al. [4] prepared Fe-based amorphous magnetic cores (AMPCs) from FeSiBCr amorphous powders with phosphate–resin hybrid coatings. The high-frequency magnetic properties of AMPCs annealed at different temperatures were systematically studied. The annealed sample exhibited the lowest hysteresis loss of about 29.6 mW/cm^3 at 800 kHz, as well as a maximum effective permeability of 36.4, which could be used in high-frequency applications of up to 3 MHz. Luo et al. [5] prepared core–shell-structured amorphous FeSiBCr@phosphate/silica powders and soft magnetic composites (SMCs) via phosphation and a sodium silicate treatment. By optimized the phosphating process, a uniform and dense insulation layer can be formed on the powder surface, which is beneficial for the subsequent coating of sodium silicate. By optimizing the sodium silicate treatment, a complete and uniform SiO_2 layer can be formed well on the phosphated powders, leading to a double-layer core—shell structure and excellent soft magnetic properties. The SMCs exhibited excellent soft magnetic properties with a permeability value of $\mu_e = 35$ and a core loss of $P_s = 368$ kW/m^3 at 50 mT/200 kHz. Xie et al. [6] fabricated FeSiCr SMCs via the sol–gel method, and an Al_2O_3/resin composite layer was employed as insulation coating; this not only effectively reduced the core loss, increased the resistivity, and improved the quality factor but also increased the thermal conductivity of the SMCs. A high thermal conductivity is beneficial to enhancing the high-temperature performance, lifetime, and reliability of SMCs.

Magnetocaloric and magnetostrictive materials have also attracted significant attention. Zhong et al. [7] prepared magnetocaloric $La(Fe,Si)_{13}$/Ce-Co composites via a one-step sintering process. Using 15 wt.% Ce_2Co_7 as a dopant and binder, the Curie temperature (T_C) increased from 212 K to 331 K, and the change in the maximum magnetic entropy $(-\Delta S_M)_{max}$ decreased from 8.8 to 6.0 J/kg·K under a 5 T field. High values of compressive strength of up to 450 MPa and high thermal conductivity values of up to 7.5 W/m·K were obtained. Their work demonstrates that the one-step sintering process is a feasible route to producing $La(Fe,Si)_{13}$-based magnetocaloric composites with large MCE values, good mechanical properties, attractive thermal conductivities, and tunable T_C values. Yang et al. [8] reviewed the recent advances in magnetostrictive Tb-Dy-Fe alloys. They began with a brief introduction to the characteristics of Tb-Dy-Fe alloys and then focused on the research progress in recent years, including improved processes such as directional solidification, the magnetic field-assisted process, ferromagnetic MPB theory and sensor applications, and the reconstruction of the grain boundary phase for sintered composite materials. This review will be helpful for the design of novel magnetostrictive Tb-Dy-Fe alloys with improved properties.

He et al. [9] carried out a quantitative study of the surface composition of a NiFe thin film exposed to atmospheric conditions via angle-resolved X-ray photoelectron spectroscopy (ARXPS). The coexistence of metallic and oxidized species on the surface was demonstrated. The thicknesses of the oxidized species, including NiO, $Ni(OH)_2$, Fe_2O_3, and Fe_3O_4, were also estimated. This work provides an effective approach to clarifying the surface composition and demonstrated the dependence between the magnetic properties and thicknesses of NiFe thin films.

Fluxgate magnetometers are commonly used to detect weak magnetic targets, but the detection accuracy of a fluxgate magnetometer is affected by its own error. To obtain more accurate detection data, the sensor must be error-corrected prior to its application. Previous researchers easily fell into the local minimum when solving error parameters. In a paper by Li et al. [10], an error correction method is proposed to tackle the problem;

this method combines the dragonfly algorithm (DA) and the Levenberg–Marquardt (LM) algorithm, thereby solving the problem of the LM algorithm and improving the accuracy of solving the error parameters. The simulation results show that the DA–LM algorithm can accurately solve the error parameters of the triaxial magnetic sensor, and the difference between the corrected and the ideal total value was decreased from 300 nT to 5 nT.

Due to the complex environment of the ocean, underwater equipment has become a very threatening means of surprise attack in modern warfare. The timely and effective detection of underwater moving targets is the key to obtaining warfare advantages and has important strategic significance for national security. Xu et al. [11] proposed a moving target detection method based on magnetic flux induction technology. Their results showed their technology has an obvious response to moving targets and can effectively capture target signals.

3. Conclusions and Outlook

This Special Issue focuses on the preparation, microstructure, and properties of various metal-containing magnetic materials, including hard magnetic materials, soft magnetic materials, magnetocaloric materials, magnetostrictive materials, magnetic thin films, and magnetic technologies. Some advanced approaches to producing and synthesizing hard magnetic and soft magnetic materials and magnetic refrigeration materials are described. The surface characterization of magnetic thin films is proposed. In addition, we also accepted contributions which focused on magnetic technologies with reduced errors or improved detection. The main goal of this Special Issue to describe the most recent advances in strong magnetic materials and emerged magnetic technologies has been fulfilled.

Hard and soft magnets are the fundamental materials for electrical and electronic devices and will be constantly developed with the aim of achieving high levels of performance, small sizes, and low costs with the help of new concepts and new technologies. Other magnetic materials based on the interactions between magnetic properties and other properties, such as thermal, electric, optical, and mechanical properties, have attracted and will attract more attention in the future.

I would like to thank all the Authors for their contributions to this Special Issue and thank the managing office of Metals (MDPI) for their constant support and hard work.

Conflicts of Interest: The author declares no conflict of interest.

References

1. He, J.; Cao, J.; Yu, Z.; Song, W.; Yu, H.; Hussain, M.; Liu, Z. Grain Boundary Diffusion Sources and Their Coating Methods for Nd-Fe-B Permanent Magnets. *Metals* **2021**, *11*, 1434. [CrossRef]
2. Xiao, C.; Zeng, W.; Tang, Y.; Lu, C.; Tang, R.; Zheng, Z.; Liao, X.; Zhou, Q. Effect of Ho Substitution on Magnetic Properties and Microstructure of Nanocrystalline Nd-Pr-Fe-B Alloys. *Metals* **2022**, *12*, 1922. [CrossRef]
3. Li, J.; Zuo, J.; Yu, H. Effects of La on Thermal Stability, Phase Formation and Magnetic Properties of Fe-Co-Ni-Si-B-La High Entropy Alloys. *Metals* **2021**, *11*, 1907. [CrossRef]
4. Yu, H.; Li, J.; Li, J.; Chen, X.; Han, G.; Tang, J.; Chen, R. Enhancing the Properties of FeSiBCr Amorphous Soft Magnetic Composites by Annealing Treatments. *Metals* **2022**, *12*, 828. [CrossRef]
5. Luo, P.; Yu, H.; Wang, C.; Yuan, H.; Liu, Z.; Wang, Y.; Yang, L.; Wu, W. Properties Optimization of Soft Magnetic Composites Based on the Amorphous Powders with Double Layer Inorganic Coating by Phosphating and Sodium Silicate Treatment. *Metals* **2023**, *13*, 560. [CrossRef]
6. Xie, Q.; Yu, H.; Yuan, H.; Han, G.; Chen, X.; Liu, Z. Enhanced Magnetic Properties and Thermal Conductivity of FeSiCr Soft Magnetic Composite with Al_2O_3 Insulation Layer Prepared by Sol-Gel Process. *Metals* **2023**, *13*, 813. [CrossRef]
7. Zhong, X.; Dong, X.; Huang, J.; Liu, C.; Zhang, H.; Huang, Y.; Yu, H.; Ramanuja, R. One-Step Sintering Process for the Production of Magnetocaloric La(Fe,Si)$_{13}$-Based Composites. *Metals* **2022**, *12*, 112. [CrossRef]
8. Yang, Z.; Li, J.; Zhou, Z.; Gong, J.; Bao, X.; Gao, X. Recent Advances in Magnetostrictive Tb-Dy-Fe Alloys. *Metals* **2022**, *12*, 341. [CrossRef]
9. He, Z.; Li, Z.; Jiang, X.; Wu, C.; Liu, Y.; Song, X.; Yu, Z.; Wang, Y.; Lan, Z.; Sun, K. Surface Investigation of $Ni_{81}Fe_{19}$ Thin Film: Using ARXPS for Thickness Estimation of Oxidation Layers. *Metals* **2021**, *11*, 2061. [CrossRef]

10. Yang, L.; Li, C.; Zhang, S.; Xu, C.; Chen, H.; Xiao, S.; Tang, X.; Li, Y. Correction Method of Three-Axis Magnetic Sensor Based on DA–LM. *Metals* **2022**, *12*, 428. [CrossRef]
11. Xu, C.; Yang, L.; Huang, K.; Gao, Y.; Zhang, S.; Gao, Y.; Meng, L.; Xiao, Q.; Liu, C.; Wang, B.; et al. Researching a Moving Target Detection Method Based on Magnetic Flux Induction Technology. *Metals* **2021**, *11*, 1967. [CrossRef]

Disclaimer/Publisher's Note: The statements, opinions and data contained in all publications are solely those of the individual author(s) and contributor(s) and not of MDPI and/or the editor(s). MDPI and/or the editor(s) disclaim responsibility for any injury to people or property resulting from any ideas, methods, instructions or products referred to in the content.

Review

Grain Boundary Diffusion Sources and Their Coating Methods for Nd-Fe-B Permanent Magnets

Jiayi He [1], Jiali Cao [1], Zhigao Yu [1], Wenyue Song [1], Hongya Yu [1], Mozaffar Hussain [2] and Zhongwu Liu [1,*]

[1] School of Materials Science and Engineering, South China University of Technology, Guangzhou 510640, China; msjiayihe@mail.scut.edu.cn (J.H.); msjlcao@mail.scut.edu.cn (J.C.); msgaozhiyu@mail.scut.edu.cn (Z.Y.); mswenyue@mail.scut.edu.cn (W.S.); yuhongya@scut.edu.cn (H.Y.)

[2] Department of Physics, Air University, PAF Complex E-9, Islamabad 44000, Pakistan; mozaffar_apc@hotmail.com

* Correspondence: zwliu@scut.edu.cn; Tel.: +86-20-2223-6906

Citation: He, J.; Cao, J.; Yu, Z.; Song, W.; Yu, H.; Hussain, M.; Liu, Z. Grain Boundary Diffusion Sources and Their Coating Methods for Nd-Fe-B Permanent Magnets. *Metals* **2021**, *11*, 1434. https://doi.org/10.3390/met11091434

Academic Editor: Sergey V. Zherebtsov

Received: 9 August 2021
Accepted: 8 September 2021
Published: 10 September 2021

Publisher's Note: MDPI stays neutral with regard to jurisdictional claims in published maps and institutional affiliations.

Copyright: © 2021 by the authors. Licensee MDPI, Basel, Switzerland. This article is an open access article distributed under the terms and conditions of the Creative Commons Attribution (CC BY) license (https://creativecommons.org/licenses/by/4.0/).

Abstract: Nd-Fe-B magnets containing no heavy rare earth (HRE) elements exhibit insufficient coercivity to withstand the demagnetization field at elevated temperatures. The grain boundary diffusion (GBD) process provides the best route to fabricate high-coercive Nd-Fe-B magnets with low consumption of expensive HRE resources. Here we give a special review on the grain boundary diffusion sources and their coating methods. Up to now, various types of grain boundary sources have been developed, starting from the earliest Tb or Dy metal. The HRE-M eutectic alloys were firstly proposed for reducing the cost of the diffusion source. After that, the diffusion sources based on light rare earth and even non rare earth elements have also been proposed, leading to new understanding of GBD. Now, the diffusion sources including inorganic compounds, metals, and alloys have been employed in the industry. At the same time, to coat the diffusion source on the magnets before diffusion treatment, various methods have been developed. Different from the previous review articles for GBD, this review gives an introduction of typical types of diffusion sources and their fabrication approaches. The effects of diffusion source on the microstructure and magnetic properties are summarized briefly. In particular, the principles and applicability of different coating approaches were discussed in detail. It is believed that this review can provide a technical guidance for the industry for designing the diffusion process and products meeting specific requirements.

Keywords: Nd-Fe-B; grain boundary diffusion; coercivity; diffusion source; coating method

1. Introduction

Nd-Fe-B permanent magnets have been widely used in various fields including conventional electric motors, renewable energy, and mobile communication industries [1–3]. The total world production of sintered Nd-Fe-B magnets in 2019 was 1.9×10^5 tons, and the demand of Nd-Fe-B magnets is constantly increasing due to the large employment of electric motors and generators in the near future [4]. The magnets in the motors and generators should operate at temperatures greater than 150 °C [3], but the Nd-Fe-B magnets without the addition of heavy rare earth (HRE) elements have insufficient coercivity (H_{cj}) to withstand the demagnetization field at high such temperatures because the hard magnetic $Nd_2Fe_{14}B$ (2:14:1 phase) compound has a low Curie point (T_c) of ~312 °C, and its anisotropy field (H_A) decreases drastically with the increasing temperature [5,6]. A conventional route for fabricating high-coercive Nd-Fe-B magnets is adding the HRE elements of Dy and Tb during smelting. However, it results in a large consumption of expensive HRE resource and a sacrifice of remanence (J_r).

The grain boundary diffusion (GBD) process for the Nd-Fe-B magnets, which was firstly proposed in 2005, provides the best route to enhance the H_{cj} with less consumption of HRE [7,8]. By this way, HRE infiltrates from the surface to the interior of the magnets during a diffusion heat treatment, mainly strengthening the surface of $Nd_2Fe_{14}B$ grains by forming (Nd,HRE)$_2$Fe$_{14}$B structured shells. With the coercivity increment of 560 kA/m, the amount of Dy introduced by GBD is only 10% of that added by the conventional route [8]. Up to now, GBD has attracted much interest from both industry and academic, and it has become an important approach for the industry to fabricate cheap yet strong products. Now, most commercial Nd-Fe-B magnets with H_{cj} > 1600 kA/m (SH grade) are fabricated by GBD [9]. Their maximum working temperatures can be greater than 150 °C.

The HRE-based compound is regarded as the first generation of diffusion source. To get rid of the dependence of HRE, in 2010, a diffusion alloy of Nd-Cu without any HRE element was demonstrated effective for coercivity enhancement, which started the research and development (R&D) of the second generation of sources based on light rare earth (LRE) elements [10]. Subsequently, in 2015, a cost-effective diffusion source of MgO was proposed [11]. It gave an idea that the non-rare earth (non-RE) compound or alloy can be used to modify the grain boundary (GB) phase as the next generation of diffusion source.

The GBD process for the Nd-Fe-B magnets have been extensively reviewed in some recently published articles [12–16]. The development of GBD sources and their positive effects on magnetic properties have been discussed in detail. Different from them, this review mainly focuses on the introduction of the design and fabrication of several typical types of GBD sources. Their coating methods of the diffusion sources are described in detail. The advantages, disadvantages, and the applied ranges of various coating methods are discussed. Since the Nd-Fe-B products are mainly fabricated under customization, different diffusion sources and coating methods can be employed to meet the specific applications. It is believed that this review can provide a technical guidance for the industry for designing the diffusion process and products.

2. Development of Diffusion Sources and Their Fabrication

Figure 1 shows a comparison of the coercivity increment after GBD by HRE, LRE, and non-RE based diffusion sources, as well as their underlying mechanisms of coercivity improvement. Among the three types of GBD sources, the HRE-based one can directly enhance the H_{cj} by increasing the H_A of 2:14:1 phase, and has been industrialized. Generally, a two-step diffusion heat treatment is needed for commercial sintered magnets. During the first step GBD, the heating temperature range is generally selected at 800 to 1000 °C to ensure that the melting GB phase provides effective diffusion channels for HRE atoms. At this stage, the surface of Nd$_2$Fe$_{14}$B grains also melts due to the eutectic reaction of Nd-Nd$_2$Fe$_{14}$B system at ~685 °C [17], which is lower than the temperature of the first step GBD. In this case, HRE atoms substitute Nd atoms in the 2:14:1 lattice at the surface of the 2:14:1 grain, forming (Nd,HRE)$_2$Fe$_{14}$B shells around the hard magnetic grains [18,19]. The temperature of the second step GBD is usually selected between 400 to 600 °C to modify the distribution of GB phase, i.e., facilitating the formation of continuous GB layers for magnetic decoupling. The reported HRE-based diffusion sources can enhance the H_{cj} by > 900 kA/m for the magnets with a thickness of <5 mm. The effective HRE containing GBD sources mainly include fluorides, hydrides, and metals/alloys [20–29].

The LRE-based alloys with low melting points can form thick and continuous GB layers, effectively isolating the hard magnetic grains for decoupling. The GBD conditions of LRE sources are similar to those of the HRE sources, i.e., using a two-step heat treatment process. At present, the effective LRE-based diffusion sources mainly include Pr- and Nd-based low-melting alloys [30–36]. The coercivity increment caused by Pr-Al-Cu reaches 700 kA/m and ~500 kA/m for 2 mm- and 10 mm-thick magnets, respectively [33]. In addition to the Pr- and Nd-based diffusion alloys, high-abundance La- and Ce-based alloys have been also studied as diffusion sources recently [34–36]. However, their caused coercivity enhancement is still marginal. Some recent researches demonstrated that the non-RE elements have positive effects on microstructure modification, i.e., wetting the GB phase and reducing the defects at 2:14:1grain/GB interfaces [11,36–39]. Therefore, various non-RE metals, alloys, and compounds have been selected as the diffusion sources. The diffusion of ZnO can lead to a coercivity enhancement of 205 kA/m in a 4-mm thick magnet [37].

So far, although the coercivity enhancement by the non-RE diffusion (<250 kA/m) is still much lower than that by the RE diffusion, the non-RE GBD is expected to improve the corrosion resistance and mechanical properties of the magnets.

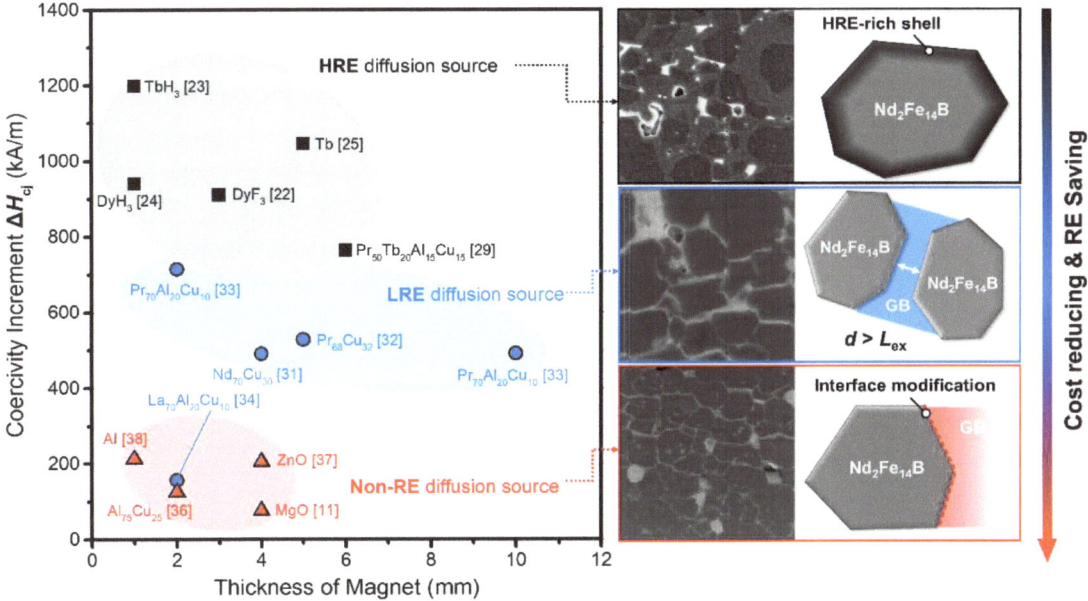

Figure 1. A comparison of coercivity increment in sintered magnets for different types of diffusion sources, including HRE, LRE, and non-RE based metals, compounds and alloys. The main mechanisms of the three generations of diffusion sources on coercivity enhancement are given.

2.1. HRE-Based Diffusion Sources and Their Fabrication

The HRE-based diffusion sources mainly include fluorides, hydrides, and metals/alloys. Generally, various rare earth (RE) oxides symbiotically exist in ores, and HRE elements should be separated from other RE elements for application. HRE oxides can be directly used for diffusion sources, but, due to their great stability and high melting points of >2000 °C, they generally exhibit poor diffusion efficiency [7,40]. Furthermore, the HRE atoms are difficult to enter into the 2:14:1 phase from the oxides, leading to a low coercivity enhancement. Therefore, the HRE oxides should be further modified to the diffusion sources with higher efficiency.

HRE fluorides are important raw materials for producing single HRE metals by thermal reduction. In the industry, the fluorides with high purity are generally fabricated from oxides by using hydrogen fluoride (HF) and ammonium hydrogen fluoride (NH_4HF_2) gases, and their reaction equations are $HRE_2O_3 + 6HF \rightarrow 2HREF_3 + 3H_2O$, and $HRE_2O_3 + 6NH_4HF_2 \rightarrow 2HREF_3 + 6NH_4F + 3H_2O$ (HRE = Dy, Tb), respectively. By using the HF gas as the reducing agent, less impurity is introduced, but, due to its high reaction temperature of 600 to 700 °C and strong causticity, it is difficult to treat the tail gas. In comparison, the reaction product from using NH_4HF_2 is easy to be recycled due to its relatively low reaction temperature of <300 °C, but it needs several repeated fluorination processes for controlling the oxygen content. DyF_3 and TbF_3 have much lower melting points of 1360 and 1172 °C, respectively, than their oxides, indicating that the HRE-F bonds have stronger tendency than HRE-O to be broken during the diffusion heat treatment at ~900 °C. This is beneficial for HRE atoms to enter into the 2:14:1 grain for improving the H_A. In addition, F^- has positive effects on saving the HRE resources. For instance, since a reaction of Nd_2O_3 +

$2DyF_3 \rightarrow (Nd,Dy)_4O_3F_6$ occurs during GBD, Dy is suppressed to be consumed at a stable $(Nd,Dy)_2O_3$ phase [41,42]. The chemical potential for Dy diffusion can be increased by F^- anions in $Nd_2Fe_{14}B$ grain, and thus the Dy atoms in the $(Nd,Dy)_4O_3F_6$ phase can diffuse into the 2:14:1 lattice [43,44].

Mass-produced HRE metals with high purity are mainly prepared by thermal reduction from their fluorides. Active non-RE metals of Ca, Mg, and Li can be employed as reductants, and Ca metal is used most widely in the industry. The reaction of $2HREF_3 + 3Ca \rightarrow 2HRE + 3CaF_2$ occurs during the reduction process. The melting points of Dy and Tb are 1409 and 1356 °C, respectively, slightly higher than their fluorides. However, it has been reported that Dy and Tb have a higher diffusion rate than their fluorides due to the different reactions occurring in the GB phase, and thus perform better in coercivity enhancement [45]. For achieving higher efficiency, the metallic HRE can be alloyed by LRE elements of Pr, Nd, La, and Ce, and non-RE elements of Al, Cu, Mg, etc., to form eutectic alloys with low melting points [26,28,32,45–47]. These added elements also play important roles in enhancing the coercivity through thickening the GBs for magnetic decoupling or reducing the defects at the interface to hinder the nucleation of reversed domains. In addition, the HRE content in the diffusion source can be reduced for reducing the material cost.

The HRE hydrides can be produced from HRE metals under a hydrogen pressure at 350 to 450 °C [48]. Compared with the HRE metals, the HRE hydride powders are more stable, indicating that the hydrides can be fabricated into the powders with smaller size. Furthermore, the hydrogen tends to be desorbed during the diffusion heat treatment. For instance, two dehydrogenation reactions of DyH_3 are $2DyH_3 \rightarrow 2DyH_2 + H_2$ (352.4 °C, $\Delta H > 0$) and $DyH_2 \rightarrow Dy + H_2$ (984.5 °C, $\Delta H > 0$) [48]. Once the hydrogen is desorbed, the powders become very reactive, which is beneficial to the diffusion of HRE. Furthermore, since a reaction of $NdO_x + xH_2 \rightarrow Nd + xH_2O$ occurs during GBD, the deoxidized Nd-rich phases have better wettability with the main phase grain, helping the formation of continuous GB layer surrounding the 2:14:1 grain for decoupling. This is also beneficial to forming the uniform HRE-rich shells [19,41,48]. However, during the GBD, the 2:14:1 could also absorb the hydrogen with a reaction of $Nd_2Fe_{14}B + (2 \pm x) H_2 \rightarrow 2NdH_{2\pm x} + 12Fe + Fe_2B + \Delta H_2$. The caused volume expansion could lead to the propagation of crack along the GB, which is not beneficial to the mechanical properties of the magnets.

Figure 2 summarizes the fabrication steps of the HRE-based diffusion sources. With the further treatment of HRE oxides, the diffusion efficiency of HRE sources can be enhanced. As a result, a higher coercivity increment can be obtained, and a thicker magnet can be treated. However, the processing cost of the diffusion sources is also increased. Therefore, the industry should select the diffusion source reasonably according to the performance requirement of the products.

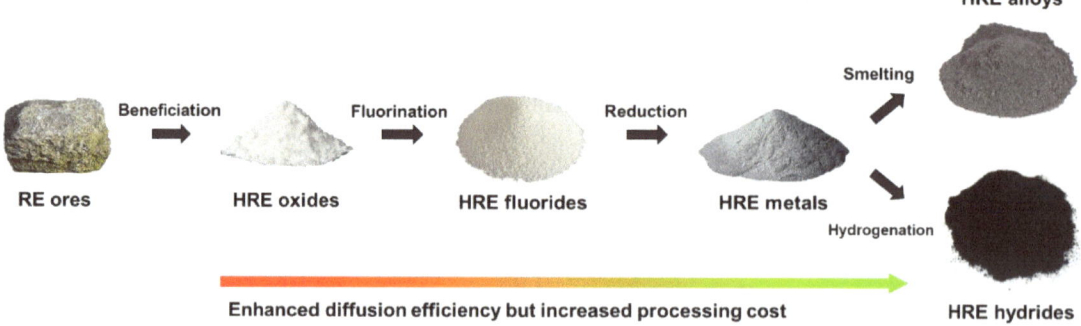

Figure 2. Fabrication steps of the HRE-based diffusion sources.

2.2. Design of LRE-Based Diffusion Sources

In comparison to the HRE-based diffusion sources, the LRE-based ones are still not widely industrialized. Due to their much lower material prices, the LRE-based diffusion sources will attract more attention in the near future and have a foreseeable development. This section mainly introduces several principles to design LRE-based diffusion alloys.

Based on the basic underlying mechanism of the LRE-based diffusion sources on coercivity improvement, most of the effective LRE sources are low-melting alloys at present. Forming LRE-M (LRE = Pr, Nd, La, Ce; M = Cu, Al, Mg, Zn, Ni) eutectic alloys by melting is a common method to obtain the low-melting diffusion sources (Figure 3a). The LRE elements, on the one hand, play significant roles in increasing the amount of the RE-rich intergranular phase for magnetic decoupling, and on the other hand, could replace Nd atoms within the 2:14:1 grains. In this case, $Pr_2Fe_{14}B$ exhibits higher H_A of 87 kOe than $Nd_2Fe_{14}B$ (67 kOe) at 300 K [6], and thus the substitution of Nd by Pr in the main phase can enhance the coercivity. In contrast, poor intrinsic magnetic properties were found in the 2:14:1 compounds of La and Ce [5,49], and the induced La and Ce are expected to segregate at GB. Previous results showed that the La, Ce-based alloys still performs much more inferior than the Pr/Nd-based ones as the diffusion source [36]. This is attributed to not only the different diffusion behavior of RE elements, but also the different wettability between the modified intergranular phase and the 2:14:1 grain.

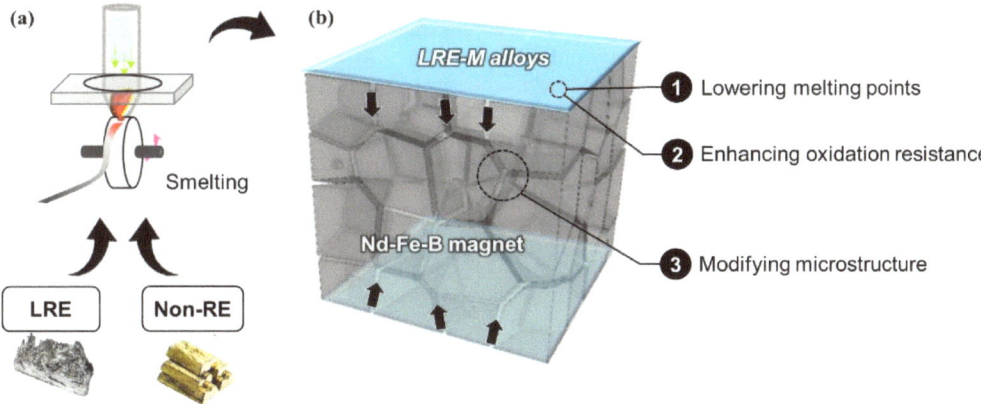

Figure 3. (**a**) Fabrication of LRE-M diffusion alloys, and (**b**) design principles in selecting non-RE alloy elements.

The non-RE elements play important roles in lowering the melting points of the diffusion source, and they also should exhibit positive effects on magnetic properties. These non-RE elements can greatly influence the diffusion behavior of LRE elements, and determine the improvement of magnetic properties at a certain extent. For instance, it was found that the coercivity enhancement caused by Pr-Al-Cu diffusion was quite sensitive to the ratio of Al/Cu [33]. The diffusion of $Pr_{70}Al_{20}Cu_{10}$ alloy can lead to an H_{cj} increase of 712 kA/m, much higher than that caused by $Pr_{70}Al_{10}Cu_{20}$ diffusion (360 kA/m). A similar phenomenon can be observed in the La-Al-Cu system [34]. Therefore, the selection of non-RE alloying elements is very important for designing effective LRE-M diffusion source. Furthermore, the non-RE elements such as Al and Ni can also modify the 2:14:1 phase by substituting Fe for a higher H_A [16]. The positive effects of the non-RE elements on microstructure modification have been summarized in a recent review article [16], which is not described in detail here.

Since the intergranular phases possess much lower corrosion potentials than the $Nd_2Fe_{14}B$ main phase, a galvanic corrosion tends to occur in corrosive medium, leading to a preferential failure of GB phases and a resultant detachment of main phase grains [50,51]. Therefore, the corrosion resistance of the magnets should be considered for practical applications, especially for those employed in seashore wind turbines. However, recent results showed that the diffusion of LRE-based alloys, such as Pr-Al-Cu and La-Al-Cu, unexpectedly decreased the corrosion resistance of the magnets due to the formed multiple intergranular phases [36]. In this case, introducing the non-RE elements with high chemical stability in GB through GBD could be a feasible route to improve the corrosion resistance of the magnets, and this deserves more attention for the future investigations. Furthermore, since LRE elements are reactive to oxygen [52], the oxidation of the alloy diffusion sources during storage and diffusion heat treatment could be a crucial reason which limits the employment of the LRE-based diffusion sources. The oxidation could severely decrease the diffusion efficiency, and particularly, deteriorate the wettability between the liquid GB phase and the 2:14:1 phase. In this case, theoretically, the non-RE elements of Al, Ni, Ti, and Cr, which exhibit a self-passivation effect in air, can be selected to alloy with the LRE to form diffusion alloys for a higher oxidation resistance. However, not many investigations have been focused on this point at the present. Much effort should be made to reveal the effects of the non-RE elements on the chemical stability of LRE-M diffusion alloys.

In summary, the design principles for LRE-M diffusion alloys in selecting non-RE alloy elements are present in Figure 3b. Firstly, the non-RE elements should be able to form eutectic alloys with the LRE elements for high diffusion efficiency. Secondly, these elements should enhance the chemical stability of the diffusion sources to avoid the oxidation during the storage and diffusion heat treatment. Finally, the added non-RE elements should also have positive effects on optimizing the microstructure. Noted that the microstructure modification is not only for enhancing the coercivity by forming continuous GB phase or reducing the amount of defects at the GB/2:14:1 interface, but also for improving the corrosion resistance of the magnets by reducing the corrosion potential between the GB and 2:14:1 phases. If the coercivity and corrosion resistance can be simultaneously enhanced by GBD process, the surface protection by anti-corrosion coatings for the Nd-Fe-B products can even be canceled, which is beneficial to further cut down the process cost. Unfortunately, the reported LRE-M diffusion sources tend to deteriorate the corrosion resistance of the magnets due to the formation of multiple RE-rich intergranular phases [36]. Therefore, the LRE-M systems should be painstakingly optimized in future investigations.

2.3. Design of Non-RE-Based Diffusion Sources

Inspired from the LRE-based diffusion alloys, if any introduced elements can modify the microstructure and wet the liquid phase to form continuous and uniform GB layers, they are candidates to be employed in the diffusion sources for the Nd-Fe-B magnets. This indicates that the critical RE elements may be not necessary for GBD. The reported non-RE based diffusion sources includes compounds and metals/alloys [34–36]. Although their positive effects on coercivity enhancement are still weaker than those of RE-based sources, it was found interesting that the diffusion of non-RE elements is effective to improve the chemical stability of the magnets [34–36]. The existing results imply that the non-RE GBD could have a broader application range than the RE GBD.

Previous investigations gave several feasible approaches to enhancing corrosion resistance of the Nd-Fe-B magnets by non-RE diffusion with different physical mechanisms. Figure 4a shows the corrosion mechanism of Nd-Fe-B magnets. Due to the strong corrosion tendency of intergranular phase, the corrosion tends to occur along the GB and corrosive media such as H_2O, O_2, and Cl^- can easily enter from the surface into the interior of the magnets through the wide corrosion channels. The GBD of non-RE oxides, including MgO [11] and ZnO [37], have been demonstrated effective to resist the corrosive medium infiltrating into the magnet. As shown in Figure 4b, the diffusion of non-RE oxides mainly lead to the formation of stable block oxides, such as Nd-Fe-O-Mg and Nd-Fe-O-Zn at triple-junction regions, which narrows the corrosion channels. The low-melting non-RE metals or alloys, such as Al [38] and $Al_{75}Cu_{25}$ [36], enhance the chemical stability of the magnets mostly by modifying the GB phases to increase their corrosion potentials (Figure 4c). These two types of non-RE sources mainly resist the corrosion process from kinetics and thermodynamics aspects, respectively, but their protection may not be as effective as that caused by an anti-corrosion coating. Since the anti-corrosion coatings can greatly isolate the reactive magnet substrate from the corrosive environment, we proposed an annealed Al-Cr coating recently for combining both surface coating and GBD [39]. Refractory Cr element was selected to modify the diffusion of Al, i.e., during the annealing, a small amount of Al was allowed to enter into the magnet for GB modification, while the added Cr led to a dense surface coating (Figure 4d). The results showed that the coercivity of the $Al_{62.5}Cr_{37.5}$ diffused magnet was increased from 1089 to 1178 kA/m. Meanwhile, in 3.0 wt.% NaCl solution, the corrosion current density of this magnet decreased significantly from 35.32 to 2.53 $\mu A/cm^2$. This method gives an idea to integrate the surface protection with the GBD process, which could further improve the competitiveness of non-RE based diffusion sources.

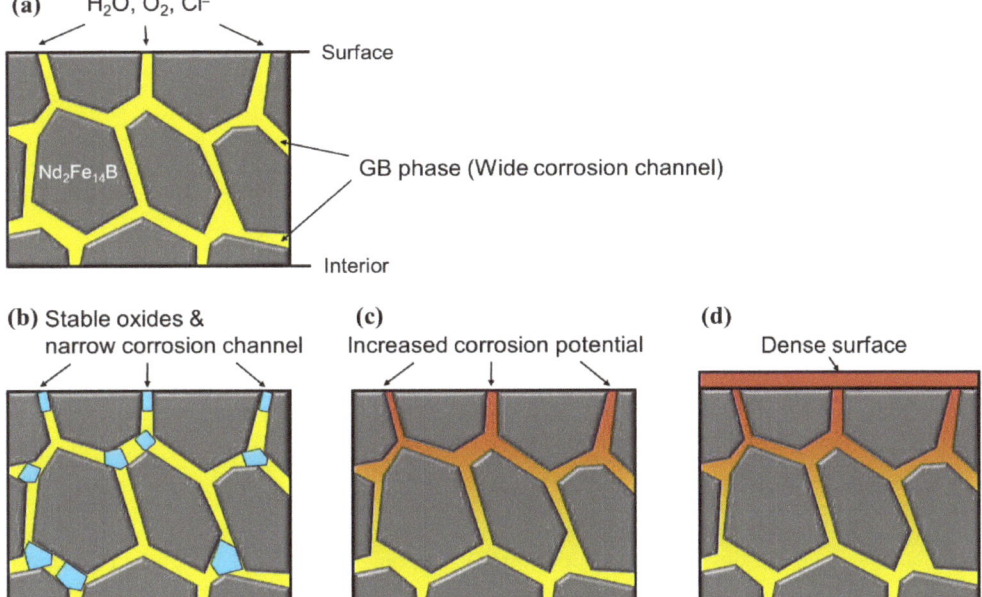

Figure 4. (**a**) The corrosion mechanism for Nd-Fe-B magnets. The schematic of improvement of corrosion resistance by GBDs by non-RE based (**b**) oxides, (**c**) low-melting metals/alloys, and (**d**) high-melting alloys.

3. Coating Methods of Diffusion Sources

Section 2 introduced various sources for GBD. Compared with the investigations of diffusion sources, the studies about how the sources can be deposited onto the magnets are relatively insufficient. However, this issue is quite critical for the industry. With the development of GBD process, more and more coating techniques have been employed for coating the diffusion sources. This section summarizes various coating methods for GBD sources and show how they have been applied or they will be employed. As shown in Figure 5, at present, the coating methods for GBD sources can be mainly classified into three types: adhesive coating, electrodeposition, and vapor deposition.

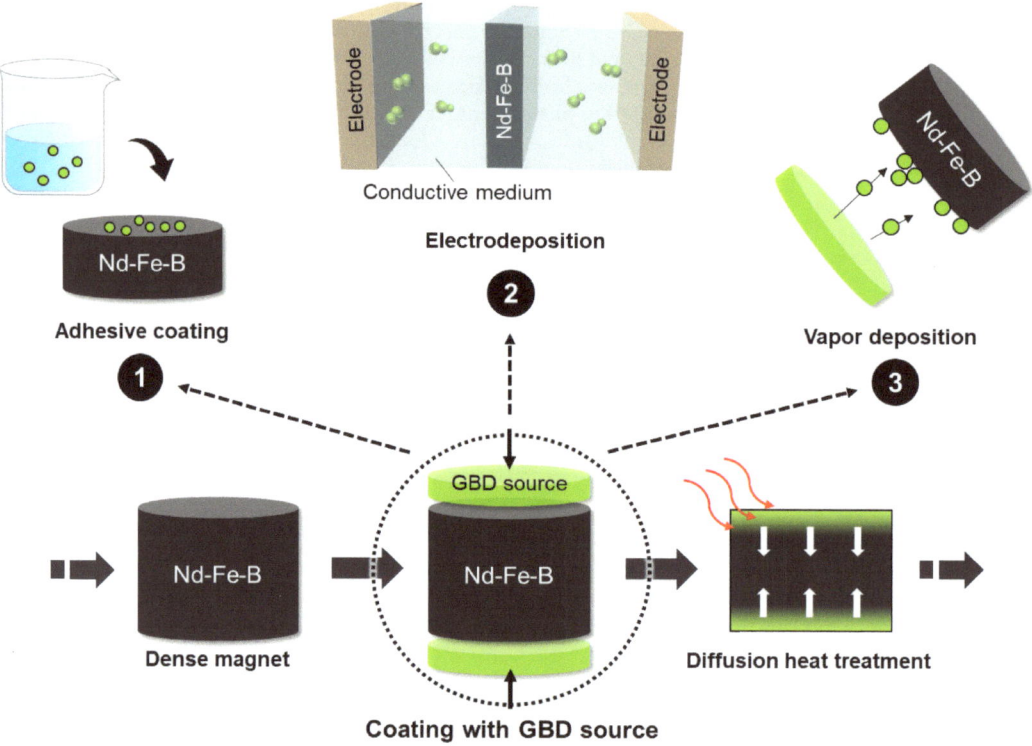

Figure 5. Processing steps of GBD and a classification of coating methods for GBD sources.

3.1. Adhesive Coating

By the adhesive coating, the powder-form diffusion sources are firstly mixed with liquid organic binders to obtain suspensions or slurries. Subsequently, the suspensions or slurries are deposited onto the surface of the magnets by dipping, spraying, and roller coating.

Dipping has been a common method to deposit the HRE inorganic compounds, including oxides [7], fluorides [7,53], and hydrides [19,41,53]. The particle size of those compounds is generally controlled at 1 to 5 μm [7,41]. By this method, the magnets are immersed in the suspensions of the diffusion source, soaked for a short time, and then removed from the container, as shown in Figure 6a. During the removal of the coated magnets, the excess diffusion sources will flow back into the container. Therefore, the dipping exhibits a high production efficiency and a simple process to deposit thick coatings with a thickness of 20 to 30 μm. However, it was suggested that the dipped coatings of diffusion sources are uneven and rough [22], indicating that it could cause an unexpected waste of HRE resource or an inhomogeneous diffusion. Furthermore, the consumption of diffusion source cannot be greatly controlled, which is not beneficial to obtaining the products with high stability.

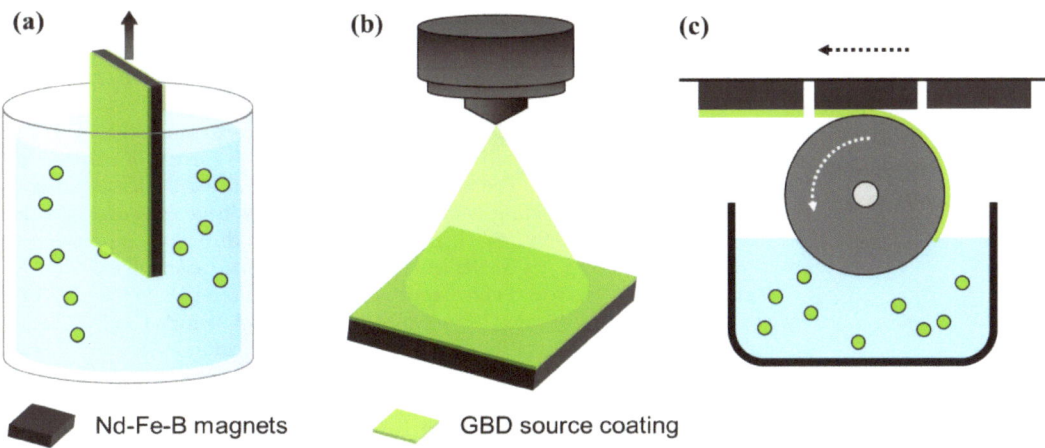

Figure 6. Schematics of (**a**) dipping, (**b**) spraying, and (**c**) roller coating methods for GBD sources.

Spraying is also effective for the inorganic compounds with small particle size of several micrometers. By the spraying, the suspensions of the diffusion source are atomized by a spray gun and deposited onto the surface of magnets, as shown in Figure 6b. This method also possesses a high production efficiency, and is suitable for both manual work and automatic production. Compared with the dipping, the spraying can easily fabricate a flat coating of diffusion source, but more diffusion sources will be consumed upon the deposition process. In addition, the highly dispersed spray can lead to a fierce volatilization of the organic solvents, which is not environmentally friendly and does harm to human health. Both the dipping and the spraying can treat profiled and large-scale magnets. For these two methods, the viscosity of the suspension greatly influences the quality of the coatings, i.e., an over-low viscosity causes an insufficient thickness of coatings while an over-high viscosity leads to a formation of uneven and over-thick coatings.

In comparison to the inorganic compounds, HRE metals/alloys are more difficult to suspend in the organic solvents due to their relatively low affinity to the organic solvents. It requires the metal/alloy powders to have a much smaller size than inorganic compound particles. However, the metal/alloy powders, especially the LRE-based alloys, are too reactive to be pulverized into fine powders. Hence, most of the reported alloy sources such as Dy-Ni-Al [3], Pr-Al [54], and Pr-Dy-Al [54], were mixed with paraffin and polyvinyl pyrrolidone (PVP), respectively, and were painted onto the surface of magnets. In this case, the roller coating provides a feasible approach to deposit the metal or alloy diffusion source carried by a roller (Figure 6c), since it is applied for not only the suspensions, but also the slurries with high viscosity. This method exhibits a simple process and is suitable to treat the large-scale magnets with regular shapes such as cube and cuboid.

In general, the adhesive coating methods exhibit a simple process and are suitable to treat magnets with large sizes. However, the amount of source coating cannot be precisely controlled. Furthermore, the organic solvents added into the diffusion sources play quite important roles to obtain an appropriate viscosity of adhesive coatings, but they have not been the focus of research. In addition, high contents of carbon [55] and oxygen [56,57] in the magnets have negative impacts on GBD since they facilitate the formation of refractory Nd-carbides and Nd-oxides, respectively, in GB. Therefore, during the temperature-rise period of diffusion heat treatment, the organic solvents should decompose into products with strong volatility to ensure the less residual carbon and oxygen on the surface of magnets. For industrialization, more attention should be focused on develop suitable solvents for various diffusion sources.

3.2. Electrodeposition

The electrodeposition for GBD sources mainly includes electroplating and electrophoresis. These two methods have quite different deposition principles, and thus exhibit different applications.

As shown in Figure 7a, by the electroplating, reduction-oxidation (REDOX) reactions occur and the metal ions in electrolytes are reduced on the surface of magnets to form metal/alloy source coatings. The magnet substrate generally acts as a cathode. In Nd-Fe-B industry, the electroplating is mainly used for depositing anti-corrosion coatings, such as Ni-P and Ni/Cu/Ni [58,59]. If the metal or alloy diffusion sources can be also prepared by electroplating, it is convenient for the companies to make full use of the existing production line. The metal and alloy source coatings, including Dy [60] and Nd-Cu [61], have been successfully fabricated by electroplating. The composition of alloy coatings can be controlled by the mixing ratio of their precursors. For instance, the Nd/Cu ratio can be regulated by changing the ratio of $Nd(NO_3)_3$ and $Cu(NO_3)_2$ in the electrolytes or the deposition potential [61]. The electroplating has a distinct advantage on fabricating smooth and thick coatings. However, the environmental concerns still exist and the pollution problems are urgent to be solved.

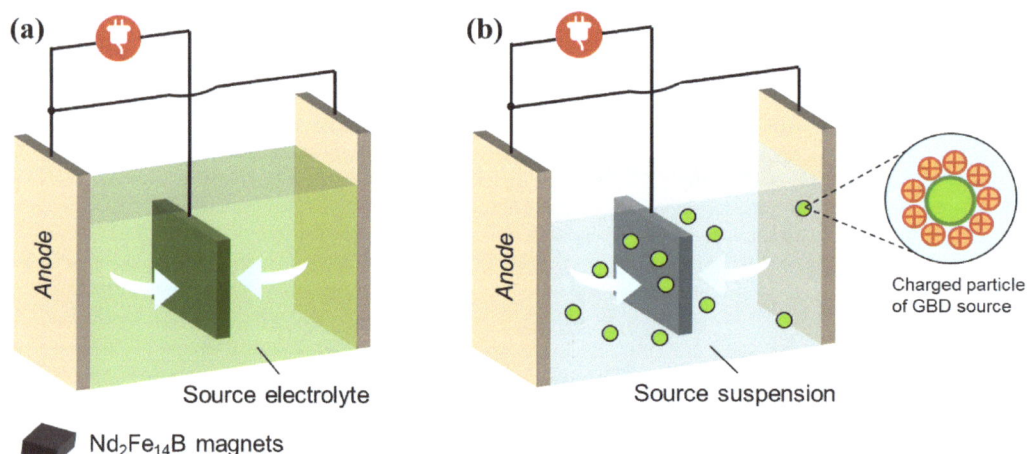

Figure 7. Schematic of electrodeposition methods for GBD sources: (**a**) electroplating and (**b**) electrophoresis. The GBD source particle is assumed to be electropositive in (**b**).

Compared with the electroplating, the electrophoresis is more applicable for GBD sources of inorganic compounds, such as Dy_2O_3 [40], DyF_3 [22,62], and TbF_3 [63,64]. As shown in Figure 7b, during the electrophoretic process, the charged particles of diffusion sources in the suspensions move towards the magnet electrodes with an opposite polarity, and the deposition can be realized. Based on the different polarity of the source particles, the magnet substrate can be anode or cathode. It was suggested that compared with the dipping, the electrophoretic deposition can fabricate the uniform coatings with lower porosity, which is beneficial to the efficient use of diffusion sources [22,64]. Under the same condition of diffusion heat treatment, the DyF_3 coatings deposited by dipping and electrophoresis enhanced the coercivity of the sintered magnets from 1200 to 1540 and 1620 kA/m, respectively [22]. However, by the electrophoretic deposition, the bonding between source particles in the coating is mainly dependent on the van der Waals force, indicating that the coatings could be easily detached from the magnets due to the poor coating/substrate adhesion. Therefore, the agglomerants such as polyethyleneimine (PEI) [40], are needed to be introduced into the suspensions.

Generally, the electrodeposition methods are applicable to treating the profiled magnets. They have advantages on fabricating dense and even coatings with thickness >10 μm, which is easy to employ for mass production. The thickness of the coating can be controlled by regulating deposition potential and keeping time. However, their processing cost is higher than that of adhesive coating methods due to their relatively longer processing cycle.

3.3. Vapor Deposition

Up to now, the vapor deposition for GBD sources is mainly physical vapor deposition (PVD). PVD is a rapidly growing technology in recent decades, which vaporizes the material source into gaseous atoms and molecules, or partially ionizes into ions by physical methods under vacuum, and realizes the deposition of films at room temperature. In comparison to the two mentioned coating methods of the adhesive coating and the electrodeposition, PVD can prepare the GBD source coatings without any effluent disposal, and therefore it can be regarded as the "greenest" coating techniques for GBD sources [65,66]. Based on the different ways to vaporize or ionize the material sources, the PVD can be mainly clarified into three types: evaporation deposition, sputtering deposition, and ion plating.

By evaporation deposition, the heated GBD sources are vaporized and sublimated onto the surface of the magnet. This method is mostly applicable for fabricating metallic and alloy coatings, and its deposition rate is relatively high among the PVD methods. Table 1 presents the temperature of several RE elementary substances for GBD in different vapor pressures. Generally, the temperature at which the saturated vapor pressure is 10^{-2} Torr is defined as the vaporization point. All of the listed RE metals for GBD exhibit the high vaporization points of >1000 °C, indicating that the costs from both equipment and energy consumption are undoubtedly increased. Previous work mainly paid attention to the evaporation of HRE metals of Dy [67] and Tb [68,69]. Among the critical RE elements of Pr, Nd, Dy, and Tb, the vaporization points of Nd and Dy are relatively low, and thus their metals and alloys are more capable for evaporation deposition.

Table 1. Temperatures of several RE metals in different vapor pressures. The data are extracted from [52].

RE Metal	Temperature in °C for a Vapor Pressure of			
	10^{-8} Atm (7.60×10^{-6} Torr)	10^{-6} Atm (7.60×10^{-4} Torr)	10^{-4} Atm (7.60×10^{-2} Torr)	10^{-2} Atm (7.60 Torr)
La	1301	1566	1938	2506
Ce	1290	1554	1926	2487
Pr	1083	1333	1701	2305
Nd	955	1175	1500	2029
Tb	1124	1354	1698	2237
Dy	804	988	1252	1685

By sputtering deposition, solid GBD source targets are bombarded by accelerated particles, and the escaped atoms or molecules reach the surface of the magnet substrates to form coatings. This approach has been widely used to deposit not only HRE metals/alloys of Dy [20,70], Tb [70], Dy-Zn [71], and Dy-Mg [72], but also non-RE diffusion sources of MgO [11], ZnO [37], and Al [38]. Compared with the evaporation, the sputtering deposition can obtain greater coating/substrate adhesion and more stable product quality. Furthermore, the thickness of coatings can be precisely controlled at a nanometer level by modifying the sputtering power and deposition time. Therefore, sputtered GBD sources is also promising to treat Nd-Fe-B films for micromechanical devices and magnetic recording media. A sputtered 50-nm thick Dy film can enhance the coercivity from 963 for a 120-nm thick Nd-Dy-Fe-B layer to 1552 kA/m, obtaining an increase of 61% [73]. However, the sputtering targets are consumable items, and in particular, the effective availability of targets are of <30%, which increase the processing cost to a certain extent.

Ion plating is a method which integrates the evaporation and the sputtering, i.e., the evaporated GBD sources can be partially ionized by gas discharge, and the ions can be accelerated by an electric field to reach the surface of the magnet substrates. This leads to a significantly improved ionization rate and, therefore, the deposition rate of ion plating is generally higher than that of the sputtering. In addition, the high ionization rate is beneficial to treat profiled magnets because the ions can move along the electric field, wrapping around the substrate and forming uniform coatings. However, the ion plating has not been widely used for depositing GBD sources, mostly due to the high evaporation temperatures of the RE metals and the caused difficulty in the equipment manufacture.

Table 2 summarizes the advantages, disadvantages, and application ranges of different coating approaches for diffusion sources. Among the mentioned coating methods, the vapor deposition has an overwhelming advantage regarding environmental protection. Meanwhile, it can precisely control the consumption of GBD sources, which is beneficial for saving the critical RE resources and improving the stability of the product. However, owing to the necessary vacuum environment and the relatively low deposition rate, the production efficiency is lower than those of the adhesive coating and the electrodeposition. Furthermore, the vapor deposition still exhibits high costs from equipment and processing. Therefore, this approach applies to small quantities of products.

Table 2. Advantages, disadvantages, and application ranges of various coating methods for depositing GBD sources.

Coating Method	Subdivision of Coating Method	Costs Advantage from Process and Equipment	Production Efficiency	Level for Materials Saving	Quality of Coating	Level of "Green"	Applicable GBD Source
Adhesive coating	Dipping	High	High	Low	Low	Moderate	Inorganic compounds
	Spraying			Low	Moderate	Low	Inorganic compounds
	Roller coating			Moderate	Low	Moderate	Inorganic compounds and metals/alloys
Electrodeposition	Electroplating	Moderate	Moderate	Moderate	High	Low	Metals/alloys
	Electrophoresis			Moderate	Moderate	Moderate	Inorganic compounds
Vapor deposition	Evaporation	Low	Low	High	High	High	Metals/alloys
	Sputtering			High	High	High	Inorganic compounds and metals/alloys
	Ion plating			High	High	High	Inorganic compounds and metals/alloys

4. Summary and Future Prospect

With the R&D of GBD process for over 20 years, the diffusion sources for Nd-Fe-B magnets have been developed for three generations of HRE, LRE, and non-RE based compounds or metals/alloys, in order to reduce the use critical RE elements and cost. At present, the GBD process can be employed to enhance not only the magnetic properties, but also other service performance such as corrosion resistance. Since the permanent magnets with less critical RE elements emerge rapidly in recent years, such as multi-main phase (MMP) magnets [54,74] and (La,Ce,Y)-Fe-B magnets [75,76], the non-RE diffusion sources are competitive to treat these cost-effective magnets. With the industrialization of GBD, various coating approaches for diffusion sources have been employed, including adhesive coating, electrodeposition, and vapor deposition. These methods have their own advantages, disadvantages, and application ranges. Since the Nd-Fe-B products are mainly fabricated under customization, the different diffusion sources and the various coating methods will be selected to meet specific applications. GBD is still in its rapid development and is far from mature. For the future investigations, more attention should be paid to develop not only efficient yet cheap diffusion sources, but also cost-effective coating methods.

Author Contributions: Conceptualization, J.H., J.C. and Z.Y.; methodology, J.H.; validation, H.Y., M.H. and Z.L.; formal analysis, J.H.; investigation, J.H.; resources, Z.L.; data curation, J.C., Z.Y. and W.S.; writing—original draft preparation, J.H.; writing—review and editing, J.H.; visualization, J.H.; supervision, Z.L.; project administration, Z.L.; funding acquisition, Z.L. All authors have read and agreed to the published version of the manuscript.

Funding: This research was funded by National Natural Science Foundation of China, grant number 51774146 and 52071143.

Institutional Review Board Statement: Not applicable.

Informed Consent Statement: Not applicable.

Data Availability Statement: Data available in a publicly accessible repository.

Acknowledgments: This work is supported by National Natural Science Foundation of China (Nos. 51774146 and 52071143).

Conflicts of Interest: The authors declare no conflict of interest.

References

1. Coey, J.M.D. Hard magnetic materials: A perspective. *IEEE Trans. Magn.* **2011**, *47*, 4671–4681. [CrossRef]
2. Gutfleisch, O.; Willard, M.A.; Bruck, E.; Chen, C.H.; Sankar, S.G.; Liu, J.P. Magnetic materials and devices for the 21st century: Stronger, lighter, and more energy efficient. *Adv. Mater.* **2011**, *23*, 821–842. [CrossRef]
3. Oono, N.; Sagawa, M.; Kasada, R.; Matsui, H.; Kimura, A. Production of thick high-performance sintered neodymium magnets by grain boundary diffusion treatment with dysprosium–nickel–aluminum alloy. *J. Magn. Magn. Mater.* **2011**, *323*, 297–300. [CrossRef]
4. Hu, B. Status quo of rare earth permanent magnet industry. In Proceedings of the Conference on Green Development and Efficient Utilization of Rare Earth Resources, Ganzhou, China, 20 October 2020.
5. Grossinger, R.; Sun, X.K.; Eibler, R.; Buschow, K.H.J.; Kirchmayr, H.R. Temperature dependence of anisotropy fields and initial susceptibilities in $R_2Fe_{14}B$ compounds. *J. Magn. Magn. Mater.* **1986**, *58*, 55–60. [CrossRef]
6. Hirosawa, S.; Matsuura, Y.; Yamamoto, H.; Fujimura, S.; Sagawa, M.; Yamauchi, H. Magnetization and magnetic anisotropy of $R_2Fe_{14}B$ measured on single crystals. *J. Appl. Phys.* **1986**, *59*, 873–879. [CrossRef]
7. Nakamura, H.; Hirota, K.; Shimao, M.; Minowa, T.; Honshima, M. Magnetic properties of extremely small Nd-Fe-B sintered magnets. *IEEE Trans. Magn.* **2005**, *41*, 3844–3846. [CrossRef]
8. Sugimoto, S. Current status and recent topics of rare-earth permanent magnets. *J. Phys. D Appl. Phys.* **2011**, *44*, 064001. [CrossRef]
9. JLMAG RARE-EARTH CO., LTD. Available online: http://www.jlmag.com.cn/view44-1.html (accessed on 7 August 2021).
10. Sepehri-Amin, H.; Ohkubo, T.; Nishiuchi, T.; Hirosawa, S.; Hono, K. Coercivity enhancement of hydrogenation–disproportionation–desorption–recombination processed Nd–Fe–B powders by the diffusion of Nd–Cu eutectic alloys. *Scr. Mater.* **2010**, *63*, 1124–1127. [CrossRef]
11. Zhou, Q.; Liu, Z.W.; Zhong, X.C.; Zhang, G.Q. Properties improvement and structural optimization of sintered NdFeB magnets by non-rare earth compound grain boundary diffusion. *Mater. Des.* **2015**, *86*, 114–120. [CrossRef]
12. Chen, F. Recent progress of grain boundary diffusion process of Nd-Fe-B magnets. *J. Magn. Magn. Mater.* **2020**, *514*, 167227. [CrossRef]
13. Lv, M.; Kong, T.; Zhang, W.; Zhu, M.; Jin, H.; Li, W.; Li, Y. Progress on modification of microstructures and magnetic properties of Nd-Fe-B magnets by the grain boundary diffusion engineering. *J. Magn. Magn. Mater.* **2021**, *517*, 167278. [CrossRef]
14. Liu, Z.; He, J. Several issues on the grain boundary diffusion process for Nd-Fe-B permanent magnets. *Acta Metall. Sin.* **2021**, *57*, 1155–1170. (in Chinese)
15. Liu, Z.; He, J.; Ramanujan, R.V. Significant progress of grain boundary diffusion process for cost-effective rare earth permanent magnets: A review. *Mater. Des.* **2021**, *209*, 110004. [CrossRef]
16. Liu, Z.; He, J.; Zhou, Q.; Huang, Y.; Jiang, Q. Development of non-rare earth grain boundary modification techniques for Nd-Fe-B permanent magnets. *J. Mater. Sci. Technol.* **2021**, *98*, 51–61. [CrossRef]
17. Seelam, U.M.R.; Ohkubo, T.; Abe, T.; Hirosawa, S.; Hono, K. Faceted shell structure in grain boundary diffusion-processed sintered Nd-Fe-B magnets. *J. Alloys. Compd.* **2014**, *617*, 884–892. [CrossRef]
18. Löewe, K.; Brombacher, C.; Katter, M.; Gutfleisch, O. Temperature-dependent Dy diffusion processes in Nd–Fe–B permanent magnets. *Acta Mater.* **2015**, *83*, 248–255. [CrossRef]
19. Kim, T.H.; Sasaki, T.T.; Koyama, T.; Fujikawa, Y.; Miwa, M.; Enokido, Y.; Ohkubo, T.; Hono, K. Formation mechanism of Tb-rich shell in grain boundary diffusion processed Nd–Fe–B sintered magnets. *Scr. Mater.* **2020**, *178*, 433–437. [CrossRef]
20. Park, K.T.; Hiraga, K.; Sagawa, M. Effect of metal-coating and consecutive heat treatment on coercivity of thin Nd-Fe-B sintered magnets. In Proceedings of the 16th Workshop on Rare Earth Permanent Magnet and their Applications, Sendai, Japan, 1 February 2000; pp. 257–264.

21. Hirota, K.; Nakamura, H.; Minowa, T.; Honshima, M. Coercivity enhancement by the grain boundary diffusion process to Nd–Fe–B sintered magnets. *IEEE Trans. Magn.* **2006**, *42*, 2909–2911. [CrossRef]
22. Soderžnik, M.; Rožman, K.Ž.; Kobe, S.; McGuiness, P. The grain-boundary diffusion process in Nd–Fe–B sintered magnets based on the electrophoretic deposition of DyF$_3$. *Intermetallics* **2012**, *23*, 158–162. [CrossRef]
23. Liu, W.Q.; Chang, C.; Yue, M.; Yang, J.S.; Zhang, D.T.; Zhang, J.X.; Liu, Y.Q. Coercivity, microstructure, and thermal stability of sintered Nd–Fe–B magnets by grain boundary diffusion with TbH$_3$ nanoparticles. *Rare Met.* **2014**, *36*, 718–722. [CrossRef]
24. Ji, W.; Liu, W.; Yue, M.; Zhang, D.; Zhang, J. Coercivity enhancement of recycled Nd-Fe-B sintered magnets by grain boundary diffusion with DyH$_3$ nano-particles. *Phys. B* **2015**, *476*, 147–149. [CrossRef]
25. Loewe, K.; Benke, D.; Kübel, C.; Lienig, T.; Skokov, K.P.; Gutfleisch, O. Grain boundary diffusion of different rare earth elements in Nd-Fe-B sintered magnets by experiment and FEM simulation. *Acta Mater.* **2017**, *124*, 421–429. [CrossRef]
26. Liu, Y.; He, J.; Yu, H.; Liu, Z.; Zhang, G. Restoring and enhancing the coercivity of waste sintered (Nd,Ce,Gd)FeB magnets by direct Pr-Tb-Cu grain boundary diffusion. *Appl. Phys. A* **2020**, *126*, 657. [CrossRef]
27. Liu, Y.; Liao, X.; He, J.; Yu, H.; Zhong, X.; Zhou, Q.; Liu, Z. Magnetic properties and microstructure evolution of in-situ Tb-Cu diffusion treated hot-deformed Nd-Fe-B magnets. *J. Magn. Magn. Mater.* **2020**, *504*, 166685. [CrossRef]
28. Lu, K.; Bao, X.; Tang, M.; Chen, G.; Mu, X.; Li, J.; Gao, X. Boundary optimization and coercivity enhancement of high (BH)$_{max}$ Nd-Fe-B magnet by diffusing Pr-Tb-Cu-Al alloys. *Scr. Mater.* **2017**, *138*, 83–87. [CrossRef]
29. Chen, G.; Bao, X.; Lu, K.; Lv, X.; Ding, Y.; Zhang, M.; Wang, C.; Gao, X. Microstructure and magnetic properties of Nd-Fe-B sintered magnet by diffusing Pr-Cu-Al and Pr-Tb-Cu-Al alloys. *J. Magn. Magn. Mater.* **2019**, *477*, 17–21. [CrossRef]
30. Akiya, T.; Liu, J.; Sepehri-Amin, H.; Ohkubo, T.; Hioki, K.; Hattori, A.; Hono, K. Low temperature diffusion process using rare earth-Cu eutectic alloys for hot-deformed Nd-Fe-B bulk magnets. *J. Appl. Phys.* **2014**, *115*, 17a766. [CrossRef]
31. Chen, F.; Zhang, T.; Wang, J.; Zhang, L.; Zhou, G. Coercivity enhancement of a Nd–Fe–B sintered magnet by diffusion of Nd$_{70}$Cu$_{30}$ alloy under pressure. *Scr. Mater.* **2015**, *107*, 38–41. [CrossRef]
32. Lu, K.; Bao, X.; Tang, M.; Sun, L.; Li, J.; Gao, X. Influence of annealing on microstructural and magnetic properties of Nd-Fe-B magnets by grain boundary diffusion with Pr-Cu and Dy-Cu alloys. *J. Magn. Magn. Mater.* **2017**, *441*, 517–522. [CrossRef]
33. Zeng, H.; Liu, Z.; Li, W.; Zhang, J.; Zhao, L.; Zhong, X.; Yu, H.; Guo, B. Significantly enhancing the coercivity of NdFeB magnets by ternary Pr-Al-Cu alloys diffusion and understanding the elements diffusion behavior. *J. Magn. Magn. Mater.* **2019**, *471*, 97–104. [CrossRef]
34. Zeng, H.X.; Wang, Q.X.; Zhang, J.S.; Liao, X.F.; Zhong, X.C.; Yu, H.Y.; Liu, Z.W. Grain boundary diffusion treatment of sintered NdFeB magnets by low cost La-Al-Cu alloys with various Al/Cu ratios. *J. Magn. Magn. Mater.* **2019**, *490*, 165498. [CrossRef]
35. Zeng, H.X.; Yu, H.Y.; Zhou, Q.; Zhang, J.S.; Liao, X.F.; Liu, Z.W. Clarifying the effects of La and Ce in the grain boundary diffusion sources on sintered NdFeB magnets. *Mater. Res. Express* **2019**, *6*, 106105. [CrossRef]
36. Zeng, H.X.; Liu, Z.W.; Zhang, J.S.; Liao, X.F.; Yu, H.Y. Towards the diffusion source cost reduction for NdFeB grain boundary diffusion process. *J. Mater. Sci. Technol.* **2020**, *36*, 50–54. [CrossRef]
37. Wang, E.; Xiao, C.; He, J.; Lu, C.; Hussain, M.; Tang, R.; Zhou, Q.; Liu, Z. Grain boundary modification and properties enhancement of sintered Nd-Fe-B magnets by ZnO solid diffusion. *Appl. Surf. Sci.* **2021**, *565*, 150545. [CrossRef]
38. Chen, W.; Huang, Y.L.; Luo, J.M.; Hou, Y.H.; Ge, X.J.; Guan, Y.W.; Liu, Z.W.; Zhong, Z.C.; Wang, G.P. Microstructure and improved properties of sintered Nd-Fe-B magnets by grain boundary diffusion of non-rare earth. *J. Magn. Magn. Mater.* **2019**, *476*, 134–141. [CrossRef]
39. He, J.; Liao, X.; Lan, X.; Qiu, W.; Yu, H.; Zhang, J.; Fan, W.; Zhong, X.; Liu, Z. Annealed Al-Cr coating: A hard anti-corrosion coating with grain boundary modification effect for Nd-Fe-B magnets. *J. Alloys. Compd.* **2021**, *870*, 159229. [CrossRef]
40. Guan, Y.W.; Huang, Y.L.; Rao, Q.; Li, W.; Hou, Y.H.; Luo, J.M.; Pang, Z.S.; Mao, H.Y. Investigation on the grain boundary diffusion of Dy$_2$O$_3$ film prepared by electrophoretic deposition for sintered Nd-Fe-B magnets. *J. Alloys. Compd.* **2020**, *857*, 157606. [CrossRef]
41. Kim, T.H.; Lee, S.R.; Kim, H.J.; Lee, M.W.; Jang, T.S. Simultaneous application of Dy-X (X = F or H) powder doping and dip-coating processes to Nd–Fe–B sintered magnets. *Acta Mater.* **2015**, *93*, 95–104. [CrossRef]
42. Takashima, M.; Kano, G. Preparation and electrical conductivity of binary rare earth metal fluoride oxides. *Solid State Ionics Diffus. React.* **1987**, *23*, 99–106. [CrossRef]
43. Kim, T.H.; Lee, S.R.; Kim, H.J.; Lee, M.W.; Jang, T.S. Magnetic and microstructural modification of the Nd-Fe-B sintered magnet by mixed DyF$_3$/DyH$_x$ powder doping. *J. Appl. Phys.* **2014**, *115*, 17A763. [CrossRef]
44. Park, S.E.; Kim, T.H.; Lee, S.R.; Kim, D.H.; Nam-Kung, S.; Jang, T.S. Magnetic and microstructural characteristics of Nd-Fe-B sintered magnets doped with Dy$_2$O$_3$ and DyF$_3$ powders. *IEEE Trans. Magn.* **2011**, *47*, 3259–3262. [CrossRef]
45. Li, D.; Suzuki, S.; Kawasaki, T.; Machida, K.I. Grain interface modification and magnetic properties of Nd–Fe–B sintered magnets. *Japn. J. Appl. Phys.* **2008**, *47*, 7876–7878. [CrossRef]
46. Wong, Y.J.; Chang, H.W.; Lee, Y.I.; Chang, W.C.; Chiu, C.H.; Mo, C.C. Comparison on the coercivity enhancement of sintered NdFeB magnets by grain boundary diffusion with low-melting (Tb, R)$_{75}$Cu$_{25}$ alloys (R= None, Y, La, and Ce). *AIP Adv.* **2019**, *9*, 125238. [CrossRef]
47. Wong, Y.J.; Chang, H.W.; Lee, Y.I.; Chang, W.C.; Chiu, C.H.; Mo, C.C. Coercivity enhancement of thicker sintered NdFeB magnets by grain boundary diffusion with low-melting Tb$_{75-x}$Ce$_x$Cu$_{25}$ (x = 0–45) alloys. *J. Magn. Magn. Mater.* **2020**, *515*, 167287. [CrossRef]

48. Liu, P.; Ma, T.; Wang, X.; Zhang, Y.; Yan, M. Role of hydrogen in Nd-Fe-B sintered magnets with DyH$_x$ addition. *J. Alloys. Compd.* **2015**, *628*, 282–286. [CrossRef]
49. Yan, M.; Jin, J.; Ma, T. Grain boundary restructuring and La/Ce/Y application in Nd–Fe–B magnets. *Chin. Phys. B* **2019**, *28*, 077507. [CrossRef]
50. Wu, Y.; Gao, Z.; Xu, G.; Liu, J.; Xuan, H.; Liu, Y.; Yi, X.; Chen, J.; Han, P. Current status and challenges in corrosion and protection strategies for sintered NdFeB magnets. *Acta Metall. Sin.* **2020**, *57*, 171–181. (In Chinese)
51. Peng, B.; Jin, J.; Liu, Y.; Lu, C.; Li, L.; Yan, M. Towards peculiar corrosion behavior of multi-main-phase Nd-Ce-Y-Fe-B permanent material with heterogeneous microstructure. *Corros. Sci.* **2020**, *177*, 108972. [CrossRef]
52. Gschneidner, K.A.; Eyring, L.; Hüfner, S. *Handbook on the Physics and Chemistry of Rare Earths*, 4th ed.; North Holland: North Holland, The Netherlands, 1994.
53. Bae, K.H.; Kim, T.H.; Lee, S.R.; Kim, H.J.; Lee, M.W.; Jang, T.S. Magnetic and microstructural characteristics of DyF$_3$/DyH$_x$ dip-coated Nd–Fe–B sintered magnets. *J. Alloys. Compd.* **2014**, *612*, 183–188. [CrossRef]
54. Jin, J.; Chen, W.; Li, M.; Liu, X.; Yan, M. PrAl and PrDyAl diffusion into Nd-La-Ce-Fe-B sintered magnets: Critical role of surface microstructure in the magnetic performance. *Appl. Surf. Sci.* **2020**, *529*, 147028. [CrossRef]
55. Sasaki, T.T.; Ohkubo, T.; Une, Y.; Kubo, H.; Sagawa, M.; Hono, K. Effect of carbon on the coercivity and microstructure in fine-grained Nd–Fe–B sintered magnet. *Acta Mater.* **2015**, *84*, 506–514. [CrossRef]
56. Mo, W.; Zhang, L.; Liu, Q.; Shan, A.; Wu, J.; Komuro, M. Dependence of the crystal structure of the Nd-rich phase on oxygen content in an Nd–Fe–B sintered magnet. *Scr. Mater.* **2008**, *59*, 179–182. [CrossRef]
57. Bae, K.H.; Lee, S.R.; Kim, H.J.; Lee, M.W.; Jang, T.S. Effect of oxygen content of Nd–Fe–B sintered magnet on grain boundary diffusion process of DyH$_2$ dip-coating. *J. Appl. Phys.* **2015**, *118*, 203902. [CrossRef]
58. Minowa, T.; Yoshikawa, M.; Honshima, M. Improvement of the corrosion resistance on Nd-Fe-B magnet with nickel plating. *IEEE Trans. Magn.* **1989**, *25*, 3776–3778. [CrossRef]
59. Cheng, C.W.; Man, H.C.; Cheng, F.T. Magnetic and corrosion characteristics of Nd-Fe-B magnet with various surface coatings. *IEEE Trans. Magn.* **1997**, *33*, 3910–3912. [CrossRef]
60. Tang, X.T.; Lu, Z.W.; Sun, A.Z. The effect of sintered Nd-Fe-B with Dy infiltration to the plating crafts. *J. Magn. Magn. Mater.* **2019**, *475*, 10–13. [CrossRef]
61. Lee, S.; Kwon, J.; Cha, H.-R.; Kim, K.M.; Kwon, H.W.; Lee, J.; Lee, D. Enhancement of coercivity in sintered Nd-Fe-B magnets by grain-boundary diffusion of electrodeposited Cu-Nd Alloys. *Met. Mater. Int.* **2016**, *22*, 340–344. [CrossRef]
62. Cao, X.J.; Chen, L.; Guo, S.; Li, X.B.; Yi, P.P.; Yan, A.R.; Yan, G.L. Coercivity enhancement of sintered Nd–Fe–B magnets by efficiently diffusing DyF$_3$ based on electrophoretic deposition. *J. Alloys. Compd.* **2015**, *631*, 315–320. [CrossRef]
63. Cao, X.; Chen, L.; Guo, S.; Chen, R.; Yan, G.; Yan, A. Impact of TbF$_3$ diffusion on coercivity and microstructure in sintered Nd–Fe–B magnets by electrophoretic deposition. *Scr. Mater.* **2016**, *116*, 40–43. [CrossRef]
64. Soderžnik, M.; Korent, M.; Žagar Soderžnik, K.; Katter, M.; Üstüner, K.; Kobe, S. High-coercivity Nd-Fe-B magnets obtained with the electrophoretic deposition of submicron TbF$_3$ followed by the grain-boundary diffusion process. *Acta Mater.* **2016**, *115*, 278–284. [CrossRef]
65. Navinšek, B.; Panjan, P.; Milošev, I. PVD coatings as an environmentally clean alternative to electroplating and electroless processes. *Surf. Coat. Technol.* **1999**, *116–119*, 476–487. [CrossRef]
66. He, J.; Lan, X.; Wan, J.; Liu, H.; Liu, Z.; Jiao, D.; Zhong, X.; Cheng, Y.; Qiu, W. Modifying Cr/CrN composite structure by Fe addition: Toward manufacturing cost-effective and tough hard coatings. *Appl. Surf. Sci.* **2021**, *545*, 149025. [CrossRef]
67. Sepehri-Amin, H.; Ohkubo, T.; Hono, K. The mechanism of coercivity enhancement by the grain boundary diffusion process of Nd–Fe–B sintered magnets. *Acta Mater.* **2013**, *61*, 1982–1990. [CrossRef]
68. Watanabe, N.; Itakura, M.; Kuwano, N.; Li, D.; Suzuki, S.; Machida, K.I. Microstructure analysis of sintered Nd-Fe-B magnets improved by Tb-vapor sorption. *Mater. Trans.* **2007**, *48*, 915–918. [CrossRef]
69. Watanabe, N.; Umemoto, H.; Ishimaru, M.; Itakura, M.; Nishida, M.; Machida, K. Microstructure analysis of Nd-Fe-B sintered magnets improved by Tb-metal vapour sorption. *J. Micros.* **2009**, *236*, 104–108. [CrossRef]
70. Matchida, K.; Suzuki, S.; Kawasaki, T.; Li, D.S.; Kitamon, T.; Nakamura, K.; Shimizu, Y. High-coercive Nd-Fe-B sintered magnets diffused with Dy or Tb metal and their applications. In Proceeding of the INTERMAG Asia 2005: Digest of the IEEE International Magnetics Conference, Nagoya, Japan, 4–8 April 2005; pp. 947–948.
71. Li, J.; Guo, C.; Zhou, T.; Qi, Z.; Yu, X.; Yang, B.; Zhu, M. Effects of diffusing DyZn film on magnetic properties and thermal stability of sintered NdFeB magnets. *J. Magn. Magn. Mater.* **2018**, *454*, 215–220. [CrossRef]
72. Zhong, S.; Munan, Y.; Rehman, S.U.; Yaojun, L.; Jiajie, L.; Yang, B. Microstructure, magnetic properties and diffusion mechanism of DyMg co-deposited sintered Nd-Fe-B magnets. *J. Alloys. Compd.* **2020**, *819*, 153002. [CrossRef]
73. Gong, W.J.; Wang, X.; Liu, W.; Guo, S.; Wang, Z.H.; Cui, W.B.; Zhu, Y.L.; Zhang, Y.Q.; Zhang, Z.D. Enhancing the perpendicular anisotropy of NdDyFeB films by Dy diffusion process. *J. Appl. Phys.* **2012**, *111*, 07a729. [CrossRef]
74. Jin, J.; Yan, M.; Liu, Y.; Peng, B.; Bai, G. Attaining high magnetic performance in as-sintered multi-main-phase Nd-La-Ce-Fe-B magnets: Toward skipping the post-sinter annealing treatment. *Acta Mater.* **2019**, *169*, 248–259. [CrossRef]

75. Zhang, J.S.; Liao, X.F.; Xu, K.; He, J.Y.; Fan, W.B.; Yu, H.Y.; Zhong, X.C.; Liu, Z.W. Enhancement in hard magnetic properties of nanocrystalline (Ce,Y)–Fe–Si–B alloys due to microstructure evolution caused by chemical heterogeneity. *J. Mater. Chem. C* **2020**, *8*, 14855–14863. [CrossRef]
76. Liao, X.; Zhang, J.; He, J.; Fan, W.; Yu, H.; Zhong, X.; Liu, Z. Development of cost-effective nanocrystalline multi-component (Ce,La,Y)-Fe-B permanent magnetic alloys containing no critical rare earth elements of Dy, Tb, Pr and Nd. *J. Mater. Sci. Technol.* **2021**, *76*, 215–221. [CrossRef]

Article

Effect of Ho Substitution on Magnetic Properties and Microstructure of Nanocrystalline Nd-Pr-Fe-B Alloys

Caihai Xiao [1,2,†], Weiwei Zeng [2,†], Yongli Tang [3], Cifu Lu [2], Renheng Tang [2], Zhigang Zheng [1,*], Xuefeng Liao [2] and Qing Zhou [2,*]

1. School of Materials Science and Engineering, South China University of Technology, Guangzhou 510640, China
2. Guangdong Provincial Key Laboratory of Rare Earth Development and Application, Institute of Resources Utilization and Rare Earth Development, Guangdong Academy of Sciences, Guangzhou 510640, China
3. Huizhou Feller Magnets Co., Ltd., Huizhou 518055, China
* Correspondence: mszgzheng@scut.edu.cn (Z.Z.); zqwork@grre.gd.cn (Q.Z.)
† These authors contributed equally to this work.

Abstract: The inevitable thermal demagnetization of magnets at high-temperatures is a key issue for Nd-Fe-B based permanent magnetic materials, especially for electric motors. Here, we report the effect of partially substituting the element Holmium (Ho) on the magnetic properties and microstructure of nanocrystalline melt-spun [$(NdPr)_{1-x}Ho_x$]$_{14.3}Fe_{76.9}B_{5.9}M_{2.9}$ (x = 0–0.6; M = Co, Cu, Al and Ga) alloys. It shows that Ho can enter into the main phase and significantly enhance the coercivity (H_{cj}). A large coercivity of 23.9 kOe is achieved in the x = 0.3 alloy, and the remanent magnetization (M_r) remains in balance. The abnormal elevated temperature behavior of M_r is observed in the alloys with a high amount of Ho substitution, in which the M_r of the x = 0.6 alloy increases with rising temperature from 300 K to 375 K owing to the antiparallel coupling between Ho and Fe moments. As a result, the positive value (0.050%/K) of temperature coefficient α of M_r is achieved in the x = 0.6 alloy within the temperature range of 300–400 K, in excess of that of existing Nd-Fe-B magnets. The temperature coefficient β of H_{cj} is also improved by Ho substitution, indicating the introduction of Ho in Nd-Fe-B magnets is beneficial for thermal stability. The microstructure observation of x = 0, 0.3 and 0.6 alloys confirmed the grain refinement by Ho substitution, and Ho prefers to remain in the 2:14:1 phase than Nd and Pr. The present finding provides an important reference for the efficient improvement of the thermal stability of Nd-Fe-B-type materials.

Keywords: Nd-Fe-B ribbons; Ho-substituted; coercivity; remanent magnetization; thermal stability

Citation: Xiao, C.; Zeng, W.; Tang, Y.; Lu, C.; Tang, R.; Zheng, Z.; Liao, X.; Zhou, Q. Effect of Ho Substitution on Magnetic Properties and Microstructure of Nanocrystalline Nd-Pr-Fe-B Alloys. *Metals* **2022**, *12*, 1922. https://doi.org/10.3390/met12111922

Academic Editor: Imre Bakonyi

Received: 12 October 2022
Accepted: 4 November 2022
Published: 9 November 2022

Publisher's Note: MDPI stays neutral with regard to jurisdictional claims in published maps and institutional affiliations.

Copyright: © 2022 by the authors. Licensee MDPI, Basel, Switzerland. This article is an open access article distributed under the terms and conditions of the Creative Commons Attribution (CC BY) license (https://creativecommons.org/licenses/by/4.0/).

1. Introduction

Nd-Fe-B-based magnets have been employed as the high-performance permanent magnetic material for numerous applications, such as in traction motors and generators, due to their excellent hard magnetic properties [1,2]. As required for the developing motor market, Nd-Fe-B magnets with superior thermal stability are seriously needed to maintain performance in high temperatures [3]. Enhancing coercivity is considered to be an effective way of withstanding the demagnetizing field at high service temperatures [4,5]. The heavy rare earth (RE) elements Dy and Tb are generally added to Nd-Fe-B magnets to improve their coercivity by the alloying or grain boundary diffusion method, since $Dy_2Fe_{14}B$ and $Tb_2Fe_{14}B$ exhibit higher anisotropy fields (H_A) than that of the $Nd_2Fe_{14}B$ compound [6]. However, the addition of Dy or Tb causes a problem regarding the price of the magnets, due to the limited natural abundance and the high cost of Dy/Tb. In the last few years, many efforts have been dedicated to the development of Dy/Tb-free Nd-Fe-B magnets with high coercivity to resist thermal demagnetization [3,7,8]. However, the outcomes are still insufficient.

Holmium (Ho) is one of the less-abundant heavy RE elements in the earth's crust, but its price is relatively low because it is also one of the less-frequently used. Although $Ho_2Fe_{14}B$ only presents a slightly higher H_A (75 kOe) than 73 kOe of $Nd_2Fe_{14}B$, several researchers have indicated that introducing Ho into Nd-Fe-B magnets can dramatically improve the coercivity [9]. A high coercivity of 18 kOe was realized in Dy-free sintered Nd-Fe-B magnets by the intergranular addition of Ho-Fe powder [10]. In addition, the $Ho_2Fe_{14}B$ compound shows a different temperature-dependence behavior of magnetization (M_s) from $Nd_2Fe_{14}B$, in which the M_s decreases with decreasing temperature, owing to the antiparallel coupling between Ho and Fe moments [11]. Therefore, the Ho-containing sintered Nd-Fe-B magnets exhibit superior thermal stability to that of Ho-free magnets. For nanocrystalline melt-spun Nd-Fe-B alloys, it is found that 20% Ho substitution can significantly improve both coercivity and high-temperature performance [12]. Recent work on high-abundance Ce-based melt-spun alloys shows that the coercivity of $(Ce_{1-x}Ho_x)_{14}Fe_{80}B_6$ can be dramatically improved from 5 kOe to 17 kOe with 50% Ho addition; meanwhile the Curie temperature T_c and thermal stability are also greatly improved [13]. In addition, the introduction of trace elements has a positive effect on the magnetic properties of Nd-Fe-B magnets. Substituting the element Co for Fe can significantly increase the Curie temperature (T_c) of the $Nd_2Fe_{14}B$ phase; thus, Co is usually employed to improve the thermal stability of Nd-Fe-B magnets [14]. Although Al is not favorable to T_c, it is found to be beneficial for coercivity by partially substituting Fe for a decrease in domain wall thickness [15]. Adding Ga and Cu can regulate the grain boundary, and results in coercivity enhancement [16,17].

Therefore, Ho has the potential to improve the magnetic properties of Nd-Fe-B magnets. However, until now, there has been no systematic report about the effect of Ho substitution on melt-spun Nd-Fe-B alloys. In this study, the effects of Ho substitution on the phase constitution, magnetic properties and microstructure of the melt-spun (Nd,Pr)-Fe-B alloys were systematically investigated. The magnetic properties at room and high temperature instances of the alloys after Ho addition were discussed in detail. This work provides useful guidelines for enhancing the hard magnetic properties of (Nd,Pr)-Fe-B magnets. More importantly, it offers a practical roadmap for reducing thermal demagnetization.

2. Experimental

A series of Ho-added alloy ingots with nominal compositions of $[(Nd_{0.8}Pr_{0.2})_{1-x}Ho_x]_{14.3}Fe_{76.9}B_{5.9}M_{2.9}$ (M = Co, Cu, Al and Ga) (at. %; x = 0–0.6) were prepared by induction melting technique under Ar atmosphere. The starting materials were Nd-Pr, Ho, Fe, Fe-B, Co, Cu, Al and Ga metals with purities higher than 99.9%. Hereafter, the samples are simply labeled as NPHFBM. The nanocrystalline melt-spun ribbons were prepared by melt spinning. The alloys with a total mass of 100 g were melted by induction melting, and the general steps of alloy melting were as follows: alloys were subjected to 1 kW heat preservation for 1 min for preheating, then the power was increased to 7 kW for melting; after an alloy had completely melted into alloy liquid, it was subject to heat preservation for 2 min, and then ejected onto the copper roller. The quenching rate of melt spinning was controlled by the linear speed of the copper wheel. A wheel speed of 20 m/s was applied in this work, to prepare melt-spun ribbons with a nanocrystalline structure.

The phase constitution of the samples in powder form was characterized by X-ray diffraction (XRD, D8 Advance, BRUKER, Karlsruhe, Germany) with Cu K_α radiation (λ = 1.5418 Å, 40 kV, 40 mA). The phase analysis was performed using the Rietveld refinement with GSAS software. The magnetic properties of the ribbons were measured by a Physical Property Measurement System (PPMS, EC-II, Quantum Design, San Diego, CA, USA) equipped with a vibrating sample magnetometer (VSM) at a maximum magnetic field of 50 kOe. For the magnetic measurements, the melt-spun ribbons were cut into small pieces with a length of ~5 mm and a width of ~2 mm and measured in-plane to eliminate the demagnetization effect. The microstructures were characterized by transmission electron microscopy (TEM, Tecnai G2 F20 S-TWIN, Thermo Fisher Scientific Inc., Waltham, MA,

USA) equipped with an energy dispersive spectrometer (EDS), and the specimens for TEM observation were prepared by ion milling (691, Gatan, Philadelphia, PA, USA). Ion-beam thinning was carried out from the two sides of the ribbons at an inclination angle of 8° between the beam and the specimen surface. The Nd-L_α, Pr-L_α, Ho-M_α and Fe-K_α in the EDS spectrum were selected for mapping. Average grain size and grain-size distributions were calculated by measuring the maximum diameter with the software (Nano Measurer 1.2) for N grains (N typically in the range of 100–200).

3. Results and discussion

3.1. Phase Constitution

The XRD patterns of the melt-spun NPHFBM alloys are presented in Figure 1a. The alloys with x = 0–0.3 are only composed of hard magnetic $RE_2Fe_{14}B$ (i.e., 2:14:1) phases with the tetragonal structure (space group $P4_2/mnm$). However, an additional $REFe_2$ (i.e., 1:2) Laves phase with the Cubic structure (space group $Fd\overline{3}m$) was detected with further increasing Ho substitution (x $\geq$ 0.4). $HoFe_2$ phase exhibits a Curie temperature T_c of 612 K, and thus it is ferromagnetic at room temperature. It can be speculated that the formation of this 1:2 phase would not affect the magnetization and remanence. However, the precipitation of the 1:2 phase would consume excess rare earth, resulting in the reduction of the non-magnetic RE-rich grain boundary phase. The RE-rich phase, such as Nd($dhcp$), Nd_2O_3($P\overline{3}m1$), may exist in these alloys, but it is very difficult to distinguish by XRD, due to its low content and/or its diffraction peaks overlapping. The enlarged XRD patterns (Figure 1b) within the 2θ range of 40–44° show that the characteristic diffraction peaks of the 2:14:1 phase monotonically shift to the higher angle with increasing Ho content, indicating the lattice contraction of the 2:14:1 phase. The refined lattice parameters a and c of the 2:14:1 phase are shown in Figure 1c. The a and c of the 2:14:1 phase for the Ho-free (x = 0) alloy are 8.78(9) Å and 12.22(6) Å, respectively, which monotonically decrease to 8.73(6) Å and 12.05(2) Å, respectively, with increasing x to 0.6. Ho exhibits a smaller atomic radius of 1.77 Å than that of Nd (1.83Å) and Pr (1.82 Å). Therefore, the introduction of Ho atoms into the 2:14:1 lattice would lead to a decrease in lattice parameters, resulting in lattice contraction. In addition, the diffraction peaks of the 1:2 phase also shift to a higher angle with increasing Ho content (see Figure 1b), indicating an excess of Ho atoms have entered the 1:2 lattice, which is not beneficial for the coercivity of the alloy, and also worsens the remanence. For the x = 0.4, 0.5 and 0.6 alloys, the proportion of 1:2 phase is 2.77 wt.%, 11.09 wt.% and 22.27 wt.%, respectively.

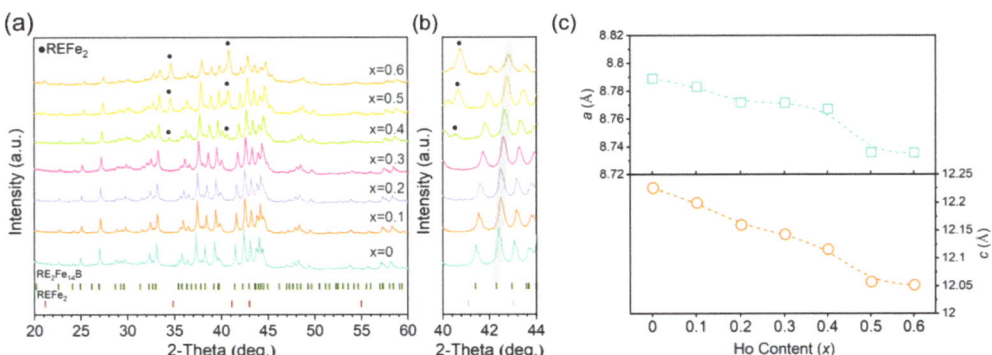

Figure 1. Phase structures of the NPHFBM alloys. (**a**) XRD patterns. (**b**) The enlarged XRD patterns within the 2θ range of 40–44°. (**c**) The refined lattice parameters of 2:14:1 phase.

3.2. Magnetic Properties

Figure 2 shows the second quadrant demagnetization curves and magnetic properties of NPHFBM alloys at 300 K. As shown in Figure 2a, the demagnetization curves of all samples present relatively good loop squareness, indicating that uniform microstructures were obtained in all alloys. No other soft magnetic phases are presented, which is also consistent with the XRD results. The intrinsic coercivity H_{cj} greatly increases from 21.1 kOe to 26.7 kOe as the doping amount (x) of Ho increases from 0 to 0.6. For the $x \geq 0.4$ alloys, the precipitation of ferromagnetic 1:2 phase is generally considered not conducive to the magnetic decoupling of the 2:14:1 main phase. However, the coercivity presents a monotonically increasing trend, which should be attributed to the higher anisotropy field H_A of the $Ho_2Fe_{14}B$ compound than that of $Nd_2Fe_{14}B$. The remanent magnetization M_r dramatically decreases from 80.7 emu/g to 29.0 emu/g with increasing x from 0 to 0.6, which could be explained by the antiferromagnetic coupling between Ho and Fe atoms. In addition, the increasing Ho content reduces the total magnetic moment per unit volume of the 2:14:1 phase, resulting in the decreasing of both saturation magnetization Ms and Mr. Here, it should be noted that the gradual growth of the coercivity shows a slowing trend compared to the linear variation of the remanence. Combined with the XRD results, it could be concluded that it is the result of the formation of the 1:2 phase instead of the 2:14:1 phase. This means that a reasonable control of Ho doping is essential to improve the overall performance of the alloy. Remarkably, the x = 0.3 alloy shows a relatively high H_{cj} of 23.9 kOe with an acceptable M_r of 55.45 emu/g, which can realize the admirable maximum magnetic energy product $(BH)_{max}$ of 48 kJ/m3.

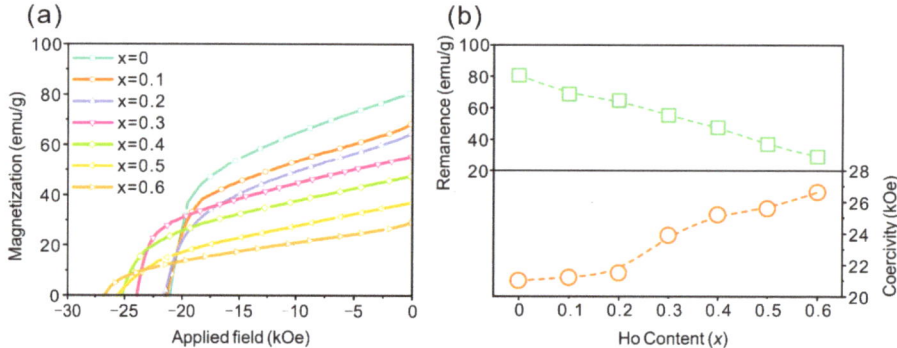

Figure 2. (a) Demagnetization curves and (b) the corresponding magnetic properties of the NPHFBM alloys at 300 K.

To fully understand the elevated temperature behavior of Ho-substituted NPHFBM alloys, the demagnetization curves of selected three alloys (x = 0, 0.3, 0.6) measured at different temperatures from 300 K to 400 K are shown in Figure 3a. The corresponding magnetic properties are presented in Figure 3b,c, respectively. As shown in Figure 3b, all selected samples show a monotonous decreasing trend of coercivity with increasing temperature from 300 K to 400 K. For the $RE_2Fe_{14}B$ compound, the exchange interaction of both RE-Fe and Fe-Fe becomes weaker with rising temperature, especially approaching the Curie temperature T_c, which also results in the decrease of H_A. Consequently, the H_{cj} decreases with increasing temperature. In addition, the x = 0.6 alloy retains 15.5 kOe of H_{cj} at 400 K, which is much higher than 11.1 kOe of the Ho-free (x = 0) alloy. It indicates that the Ho-substituted samples present higher resistance to thermal demagnetization.

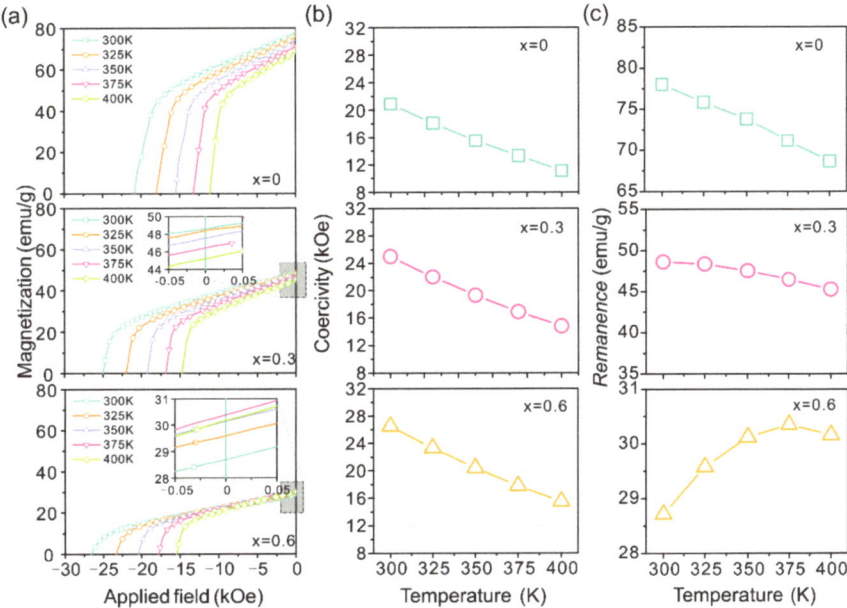

Figure 3. Temperature-dependent magnetic properties of NPHFBM alloys with x = 0, 0.3 and 0.6. (**a**) Demagnetization curves. (**b**) Coercivity. (**c**) Remanence.

Figure 3c shows the temperature dependences of M_r for x = 0, 0.3 and 0.6 alloys, respectively. The M_r for the Ho-free (x = 0) alloy decreases with increasing temperature, while the x = 0.3 alloy presents a slower downward trend. Interestingly, the M_r for the x = 0.6 alloy with high Ho content slightly increases from 28.7 emu/g to 30.4 emu/g as the temperature rises from 300 K to 375 K, and then decreases to 30.2 emu/g at 400 K. This abnormal behavior is related to the temperature-dependent magnetization M_s of the $Ho_2Fe_{14}B$ compound. As reported, the total magnetic moment (M_{tot}) of $RE_2Fe_{14}B$ is related to the coupled mode between the total RE moment (μ^{RE}) and Fe moment (μ^{Fe}) [18]. For $RE_2Fe_{14}B$ based on light RE elements, in which the RE moment is coupled parallel to the Fe moment, the total magnetic moment of $RE_2Fe_{14}B$ can be ascribed as $M_{tot} = 14\mu^{Fe} + 2\mu^{RE}$. However, for the heavy RE-based compound, antiparallel coupling occurs between the RE moment and Fe moment, which can be expressed as $M_{tot} = 14\mu^{Fe} - 2\mu^{RE}$. The spin order decreases with increasing temperature, resulting in the decrease of both μ^{Fe} and μ^{RE}. Since the reduction of the total RE moment exceeds the Fe moment, the total magnetic moment enhancement with increasing temperature can be expected, thus resulting in a slower decrease in the remanence of the alloy, or even an unexpected increase in the remanent magnetization of the alloy in a certain temperature range. However, the remanence still appears to inevitably deteriorate as the temperature increases to 400 K. This is due to the sharp decrease of the spin order at higher temperatures, resulting in the decrease of the total magnetic moment of $RE_2Fe_{14}B$.

Generally, the temperature coefficients α of remanence and β of coercivity are employed to describe the thermal stability of permanent magnetic materials, which can be defined as follows:

$$\alpha = \frac{M_r(T_2) - M_r(T_1)}{M_r(T_1) \times (T_2 - T_1)} \quad (1)$$

$$\beta = \frac{H_{cj}(T_2) - H_{cj}(T_1)}{H_{cj}(T_1) \times (T_2 - T_1)} \quad (2)$$

where T_1 is the initial temperature and T_2 is the final temperature [19]. In this work, 300 K and 400 K have been chosen as initial and final temperatures, respectively. Figure 4 presents the comparison of temperature coefficients α and β for the Ho-substituted Nd-Pr-Fe-B alloys obtained in the current work with a variety of previously reported nanocrystalline melt-spun Nd-Fe-B based alloys and commercial sintered magnets at 300–400 K. As shown by the arrow, both α and β values of the alloys obtained in this work are improved with increasing Ho doping, indicating the enhanced thermal stability of alloys with Ho substituting. Remarkably, a positive α value of 0.050 %/K is achieved in the x = 0.6 alloy. As is well known, the α value of $Nd_2Fe_{14}B$-type magnets is always negative, caused by the inverse relationship between M_s and temperature. For commercial low-end and high-end sintered Nd-Fe-B magnets, the α range is from −0.125 %/K to −0.75%/K at 300–400 K, which is slightly higher the than −0.15 ~ −0.1 %/K of reported nanocrystalline melt-spun Nd-Fe-B ribbons [12,16–18]. In addition, the Ho-substituted Nd-Fe-B alloys obtained in this work exhibit slightly higher α and β values compared with sintered Nd-Fe-B magnets, even the EH grade magnets (operating below 200 °C) with a coercivity of 30 kOe. Therefore, it suggests that the introduction of Ho into $Nd_2Fe_{14}B$-type magnets is beneficial for the magnetic properties at high temperature, especially for the M_r. This also provides a novel approach to improve the thermal stability of $Nd_2Fe_{14}B$-type magnets.

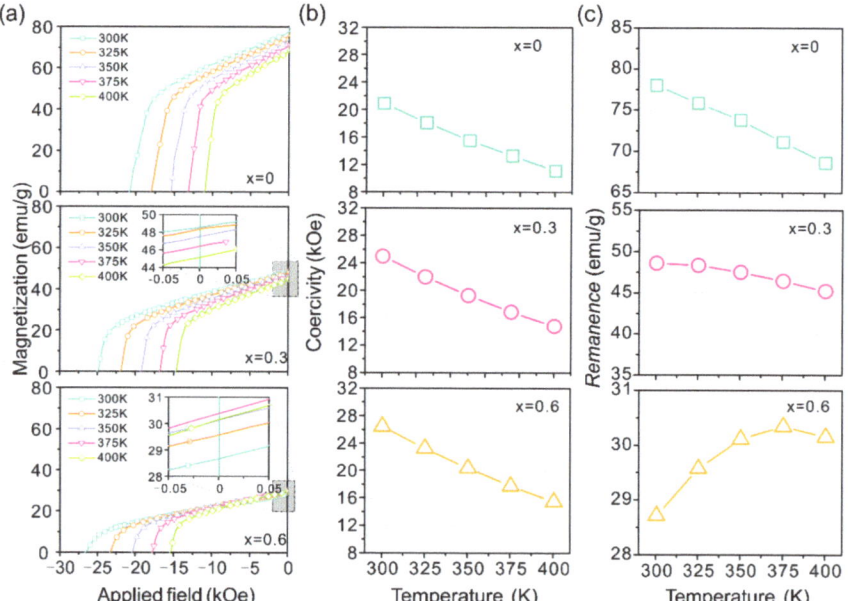

Figure 4. The comparison of temperature coefficients α and β for the Ho-substituted Nd-Fe-B alloys obtained in the current work with a variety of previously reported nanocrystalline melt-spun (ref. from [12,20,21]) and sintered Nd-Fe-B magnets (ref. from [22]) at 300–400 K. The N-, M-, H-, SH-, UH- and EH- stand for commercial sintered magnets with different coercivity ($\geq$12 kOe, $\geq$14 kOe, $\geq$17 kOe, $\geq$20 kOe, $\geq$25 kOe and $\geq$30 kOe), which is suitable for different operating temperatures.

3.3. Microstructure

Figure 5a–c shows the bright-field TEM images for the selected three alloys, i.e., x = 0, 0.3 and 0.6 alloys, respectively. The corresponding grain size distributions are presented in the inset. The well-crystallized grains have been observed in all samples. It is interesting to note that the grain sizes obtained at the same speed decrease as the amount of Ho doping increases. The Ho-free sample shows a larger mean grain size of ~72.79 nm compared with x = 0.3 alloy (~46.31 nm) and x = 0.6 alloy (~33.66 nm), which implies that Ho substitution modifies the crystallization process, which in turn affects the grain size of the alloy. As is well known, the H_{cj} for Nd-Fe-B type magnets is closely related to the grain size, and higher H_{cj} can be achieved with smaller grains [3]. The substitution of Ho for Nd/Pr can effectively refine the grains, which is beneficial for the H_{cj}. This means that the large enhancement of the coercivity is not only due to the improvement of the anisotropic field of $Ho_2Fe_{14}B$, but is also influenced by the grain size.

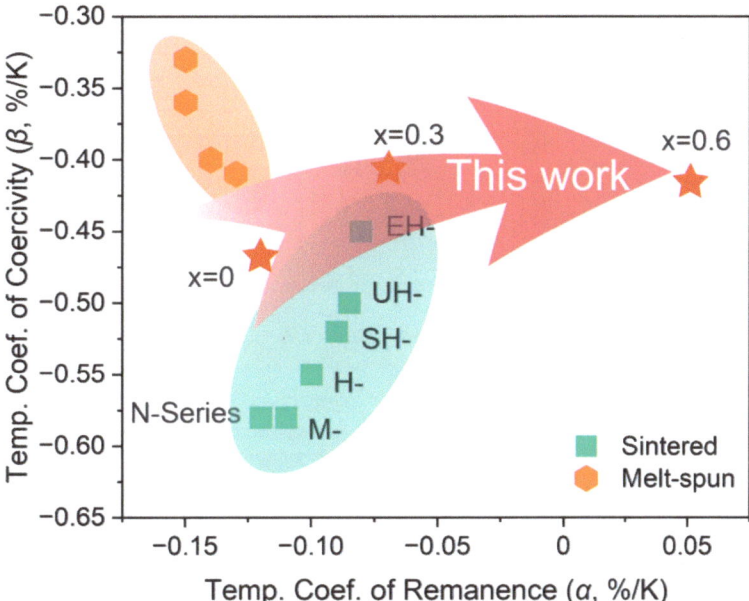

Figure 5. Microstructure characterization of NPHFBM alloys. Bright field TEM images and grain size distribution of the x = 0 (**a**), 0.3 (**b**) and 0.6 (**c**) alloys. The HRTEM image of the typical grain obtained from (**b**) for x = 0.3 alloy is shown in (**d**), and the corresponding FFT pattern is presented in the inset. The HAADF image obtained from (**b**) and corresponding EDS elemental mapping are shown in (**e**) and (**f**), respectively.

Figure 5d shows a high-resolution TEM (HRTEM) image of selected grains in Figure 5b for x = 0.3 alloy, and the corresponding fast Fourier transformation (FFT) pattern is shown as an inset in Figure 5d, confirming the grain is the 2:14:1 phase. A high-angle annular dark-field (HAADF) image obtained from Figure 5b is presented in Figure 5e, and the thin grain boundary layer with 1–2 nm is observed clearly. The GB layer plays an important role in H_{cj} improvement, in which the GB layer can isolate the magnetic coupling between the neighboring ferromagnetic 2:14:1 grains. Figure 5f presents the EDS elemental mappings taken from Figure 5e. It shows that all the RE elements are enriched in the GB layer, and Fe is mainly distributed in the 2:14:1 grains. In addition, the distribution of Ho in the GB layer is not as obvious as that of Nd and Pr, indicating the weak segregation of Ho. This is consistent with previous results, which indicated that the light RE prefers to enter into the GB compared with the heavy RE [23,24].

4. Conclusions

In this work, we systematically studied the effects of Ho addition on the phase constitution, magnetic properties and microstructures of nanocrystalline melt-spun $[(Nd_{0.8}Pr_{0.2})_{1-x}Ho_x]_{14.3}Fe_{76.9}B_{5.9}M_{2.9}$ (M=Co, Cu, Al and Ga) (at. %; x = 0–0.6) alloys. The $REFe_2$ phase is formed at high amounts of Ho substitution (x ≥ 0.4). At room temperature, the coercivity greatly improved from 21 kOe to 27 kOe with increasing Ho substitution from 0 to 60%, while the remanent magnetization deteriorated from 80.7 emu/g to 29.0 emu/g accordingly. For the elevated temperature behavior, the M_r of the x = 0.6 alloy increases with rising temperature from 300 K to 375 K, which is attributed to the antiparallel coupling between Ho and Fe moments. Both the temperature coefficient α of M_r and β of H_{cj} are improved by Ho substitution, indicating the enhanced thermal stability by Ho addition. Remarkably, the positive value of α = 0.050%/K is achieved in x = 0.6 alloy (300–400 K), which is superior to that of Nd-Fe-B magnets. The TEM results have revealed that Ho substitution can refine the size of the 2:14:1 grain. The EDS elemental mapping shows the RE segregation, in which Nd and Pr are more likely to segregate in the grain boundary phase than Ho. The present work provides a practical road map for enhancing the coercivity and thermal stability of Nd-Fe-B permanent magnetic materials simultaneously.

Author Contributions: Conceptualization, C.X. and X.L.; methodology, C.X.; software, W.Z.; validation, C.X., W.Z. and Y.T.; formal analysis, C.L.; investigation, C.X.; writing—original draft preparation, C.X.; writing—review and editing, X.L. and Q.Z..; visualization, C.L. and Y.T.; supervision, Z.Z.; funding acquisition, R.T. and Q.Z. All authors have read and agreed to the published version of the manuscript.

Funding: This work was supported by the Guangdong Basic and Applied Basic Research Foundation, China (No. 2022A1515011453, No. 2021A1515010800), and the GDAS Project of Science and Technology Development (No. 2019GDASYL-0103067, No.2022GDASZH-2022010104, No. 2022GDASZH-2022030604-04).

Conflicts of Interest: The authors declare no conflict of interest.

References

1. Coey, J.M.D. Perspective and Prospects for Rare Earth Permanent Magnets. *Engineering* **2020**, *6*, 119–131. [CrossRef]
2. Gutfleisch, O.; Willard, M.A.; Bruck, E.; Chen, C.H.; Sankar, S.G.; Liu, J.P. Magnetic materials and devices for the 21st century: Stronger, lighter, and more energy efficient. *Adv. Mater.* **2011**, *23*, 821–842. [CrossRef] [PubMed]
3. Hono, K.; Sepehri-Amin, H. Prospect for HRE-free high coercivity Nd-Fe-B permanent magnets. *Scr. Mater.* **2018**, *151*, 6–13. [CrossRef]
4. Hono, K.; Sepehri-Amin, H. Strategy for high-coercivity Nd–Fe–B magnets. *Scr. Mater.* **2012**, *67*, 530–535. [CrossRef]
5. Li, J.; Tang, X.; Sepehri-Amin, H.; Ohkubo, T.; Hioki, K.; Hattori, A.; Hono, K. On the temperature-dependent coercivities of anisotropic Nd-Fe-B magnet. *Acta Mater.* **2020**, *199*, 288–296. [CrossRef]
6. Zhou, Q.; Liu, Z.W.; Zhong, X.C.; Zhang, G.Q. Properties improvement and structural optimization of sintered NdFeB magnets by non-rare earth compound grain boundary diffusion. *Mater. Des.* **2015**, *86*, 114–120. [CrossRef]
7. Wang, E.H.; Peng, L.X.; Chen, P.Y.; Wang, M.; Zeng, W.W.; Xiao, C.H.; Lu, C.F.; Tang, R.H.; Zheng, Z.G.; Zhou, Q. Microstructure evolution and coercivity enhancement in Nd-Zn alloy-doped Nd-Fe-B powders. *J. Magn. Magn. Mater.* **2022**, *556*, 169427. [CrossRef]
8. Liu, J.; Sepehri-Amin, H.; Ohkubo, T.; Hioki, K.; Hattori, A.; Schrefl, T.; Hono, K. Grain size dependence of coercivity of hot-deformed Nd–Fe–B anisotropic magnets. *Acta Mater.* **2015**, *82*, 336–343. [CrossRef]
9. Hirosawa, S.; Matsuura, Y.; Yamamoto, H.; Fujimura, S.; Sagawa, M.; Yamauchi, H. Single Crystal Measurements of Anisotropy Constants of $R_2Fe_{14}B$ (R= Y, Ce, Pr, Nd, Gd, Tb, Dy and Ho). *Jpn. J. Appl. Phys.* **1985**, *24*, L803–L805. [CrossRef]
10. Liang, L.; Ma, T.; Wu, C.; Zhang, P.; Liu, X.; Yan, M. Coercivity enhancement of Dy-free Nd–Fe–B sintered magnets by intergranular adding $Ho_{63.4}Fe_{36.6}$ alloy. *J. Magn. Magn. Mater.* **2016**, *397*, 139–144. [CrossRef]
11. Hirosawa, S.; Matsuura, Y.; Yamamoto, H.; Fujimura, S.; Sagawa, M.; Yamauchi, H. Magnetization and Magnetic-Anisotropy of $R_2Fe_{14}B$ Measured on Single-Crystals. *J. Appl. Phys.* **1986**, *59*, 873–879. [CrossRef]
12. Brown, D.N.; Lau, D.; Chen, Z. Substitution of Nd with other rare earth elements in melt spun $Nd_2Fe_{14}B$ magnets. *AIP Adv.* **2016**, *6*, 056019. [CrossRef]
13. Li, S.; Zhou, B.; Liao, X.; Liu, Z. Tuning the hard magnetic properties of nanocrystalline Ce-Fe-B alloys by Ho substitution. *Mater. Lett.* **2022**, *323*, 132569. [CrossRef]
14. Tokunaga, M.; Nozawa, Y.; Iwasaki, K. Ga added Nd-Fe-B sintered and die-upset magnets. *IEEE Trans. Magn.* **1989**, *25*, 3561–3566. [CrossRef]
15. Strzeszewski, J.; Hadjipanayis, G.; Kim, A. The effect of Al substitution on the coercivity of Nd-Fe-B magnets. *J. Appl. Phys.* **1988**, *64*, 5568–5570. [CrossRef]
16. Fuerst, C.; Brewer, E.; Mishra, R. Die-upset Pr-Fe-B-type magnets from melt-spun ribbons. *J. Appl. Phys.* **1994**, *75*, 4208–4213. [CrossRef]
17. Kwon, H.; Yu, J. Effect of Cu-Addition and Die-Upset Temperature on Texture in Die-Upset Nd-Lean Nd-Fe-B Alloys. *J. Magn.* **2010**, *15*, 32–35. [CrossRef]

18. Coey, J.M.D.; Parkin, S.S.P. *Handbook of Magnetism and Magnetic Materials*; Springer: Cham, Switzerland, 2021.
19. Wang, E.H.; Xiao, C.H.; He, J.Y.; Lu, C.F.; Hussain, M.; Tang, R.H.; Zhou, Q.; Liu, Z.W. Grain boundary modification and properties enhancement of sintered Nd-Fe-B magnets by ZnO solid diffusion. *Appl. Surf. Sci.* **2021**, *565*, 150545. [CrossRef]
20. Chen, Z.; Wu, Y.Q.; Kramer, M.J.; Smith, B.R.; Ma, B.-M.; Huang, M.-Q. A study on the role of Nb in melt-spun nanocrystalline Nd–Fe–B magnets. *J. Magn. Magn. Mater.* **2004**, *268*, 105–113. [CrossRef]
21. Li, R.; Shang, R.X.; Xiong, J.F.; Liu, D.; Kuang, H.; Zuo, W.L.; Zhao, T.Y.; Sun, J.R.; Shen, B.G. Magnetic properties of (misch metal, Nd)-Fe-B melt-spun magnets. *AIP Adv.* **2017**, *7*, 056207. [CrossRef]
22. *Temperature Effects on Neodymium Iron Boron, Ndfeb, Magnets*. Available online: https://e-magnetsuk.com/introduction-to-neodymium-magnets/temperature-ratings/ (accessed on 20 September 2022).
23. Liao, X.F.; Zhao, L.Z.; Zhang, J.S.; Zhong, X.C.; Jiao, D.L.; Liu, Z.W. Enhanced formation of 2:14:1 phase in La-based rare earth-iron-boron permanent magnetic alloys by Nd substitution. *J. Magn. Magn. Mater.* **2018**, *464*, 31–35. [CrossRef]
24. Liao, X.F.; Zhang, J.S.; He, J.Y.; Fan, W.B.; Yu, H.Y.; Zhong, X.C.; Liu, Z.W. Development of cost-effective nanocrystalline multi-component (Ce,La,Y)-Fe-B permanent magnetic alloys containing no critical rare earth elements of Dy, Tb, Pr and Nd. *J. Mater. Sci. Technol.* **2021**, *76*, 215–221. [CrossRef]

Article

Enhanced Magnetic Properties and Thermal Conductivity of FeSiCr Soft Magnetic Composite with Al₂O₃ Insulation Layer Prepared by Sol-Gel Process

Qintian Xie [1], Hongya Yu [2,3,*], Han Yuan [2], Guangze Han [1,3,*], Xi Chen [1,3] and Zhongwu Liu [2]

[1] School of Physics and Optoelectronics, South China University of Technology, Guangzhou 510640, China
[2] School of Materials Science and Engineering, South China University of Technology, Guangzhou 510640, China
[3] Dongguan Mentech Optical & Magnetic Co., Ltd., Dongguan 523330, China
* Correspondence: yuhongya@scut.edu.cn (H.Y.); phgzhan@scut.edu.cn (G.H.)

Abstract: FeSiCr soft magnetic composites (SMCs) were fabricated by the sol-gel method, and an Al₂O₃/resin composite layer was employed as the insulation coating. By the decomposition of boehmite (AlOOH) gel into Al₂O₃ in the temperature range of 606–707 °C, a uniform Al₂O₃ layer could be formed on the FeSiCr powder surface. The Al₂O₃ insulation coating not only effectively reduced the core loss, increased the resistivity, and improved the quality factor, but it also increased the thermal conductivity of SMCs. The best overall properties with saturation magnetization M_s = 188 emu/g, effective permeability μ_e = 39, resistivity $\rho = 8.28 \times 10^5$ Ω·cm, quality factor Q = 94 at 1 MHz, and core loss = 1173 mW/cm³ at 200 kHz and 50 mT were obtained when the SMC was prepared with powders coated by 0.5 wt.% Al₂O₃ and resin. The optimized SMC exhibited the lowest core loss with 27% reduction compared to the resin only-insulated sample and 71% reduction compared to the sample without insulation treatment. Importantly, the thermal conductivity of the SMCs is 5.3 W/m·K at room temperature, which is higher than that of the samples prepared by phosphating and SiO₂ coating owing to the presence of a high thermal conductive Al₂O₃ layer. The high thermal conductivity is beneficial to enhancing the high temperature performance, lifetime, and reliability of SMCs. This work is expected to be a valuable reference for the design and fabrication of SMCs to be applied in high-temperature and high-frequency environments.

Keywords: soft magnetic composites; sol-gel; magnetic properties; core loss; thermal conductivity

Citation: Xie, Q.; Yu, H.; Yuan, H.; Han, G.; Chen, X.; Liu, Z. Enhanced Magnetic Properties and Thermal Conductivity of FeSiCr Soft Magnetic Composite with Al₂O₃ Insulation Layer Prepared by Sol-Gel Process. *Metals* **2023**, *13*, 813. https://doi.org/10.3390/met13040813

Academic Editor: Joan-Josep Suñol

Received: 27 February 2023
Revised: 16 April 2023
Accepted: 18 April 2023
Published: 21 April 2023

Copyright: © 2023 by the authors. Licensee MDPI, Basel, Switzerland. This article is an open access article distributed under the terms and conditions of the Creative Commons Attribution (CC BY) license (https://creativecommons.org/licenses/by/4.0/).

1. Introduction

As we transition to a more electrified and sustainable world, efficient power conversion requires magnetic passive devices to achieve high power density [1]. Soft magnetic materials play an important role in the field of power electronics, and they have been widely used in transformers, inductors, capacitors, etc. Soft magnetic composites (SMCs) are the newest class of soft magnetic materials, consisting of micron-sized particles pressed after insulation treatment [2]. The unique properties of SMCs include isotropic magnetic and thermal properties, high magnetic permeability, low coercivity, a high Curie temperature, reduced weight and size, and low core loss [3,4]. Hence, SMCs are capturing the attention of a growing number of researchers due to these improved electromagnetic properties [5–10].

FeSiCr alloy powder, as one of the most promising soft magnetic materials, has been employed for the preparation of SMCs [11–15]. The addition of Cr to Fe-Si alloys reduces their magnetic anisotropy and improves their corrosion resistance [16]. Insulation coatings are utilized to form the isolation network between ferrous particles so that the core loss generated by eddy currents can be reduced. The coating technology is one of the key factors for fabricating high-performance FeSiCr SMCs [17]. Both organic and inorganic materials have been used for insulation. Although the organic coatings provide satisfactory adhesion and flexibility, most of them exhibit poor resistance against high-temperature treatment [18]. Inorganic coatings usually exhibit higher thermal stability compared to organic insulations, such as epoxy and phenolic resins [19]. The latest investigations showed that the organic-inorganic composite insulation layers for FeSiCr exhibit the advantages of good adhesion and high resistivity. Guo et al. [10] improved the permeability and resistivity of FeSiCr powder by coating it with different concentrations of NiZn ferrite. Wang et al. [12] provided a new nano-$CaCO_3$/epoxy nanocomposites insulating layer on FeSiCr and carbonyl iron powder, which effectively reduced high-frequency losses. Xia et al. [17] explored the phosphating process for FeSiCr powders to improve the resistivity of SMCs. However, these studies stopped at the insulation effects, and almost no research related to the thermal conductivity of SMCs has been conducted. In fact, the SMC devices applied in AC fields are accompanied by heat generation. Excessive temperature of the magnetic core will cause a reduction in the performance of the devices or even overheating damage to the core. A high thermal conductivity of SMCs can quickly release the heat produced by magnetic losses and thus enhance the performance, lifetime, and reliability of SMCs and devices [20]. As we all know, Al_2O_3 is a metal oxide with high resistivity and temperature resistance, which is also widely used as a thermal transfer filler in polymer composites owing to its high thermal conductivity [21,22]. The thermal conductivities of typical inorganic coatings at room temperature for iron phosphate, SiO_2, Al_2O_3, and soft ferrites are 0.8–2 W/m·K [23], 0.27 W/m·K, 30 W/m·K [24], and 0.1–0.3 W/m·K [25], respectively. Al_2O_3, as an easily prepared metal oxide with both high thermal conductivity and high resistivity, is thus a good insulation layer material for SMCs. Therefore, it is expected that the magnetic performance and thermal stability of SMCs can be enhanced by an Al_2O_3 inorganic layer.

In this work, an Al_2O_3 insulation layer was successfully prepared on the surface of FeSiCr powders using the sol-gel method. The thickness of the Al_2O_3 insulation layer was controlled by the solid content of boehmite (AlOOH) sol. The microstructure and magnetic properties of the SMCs with insulation layers containing different Al_2O_3 contents were investigated. The results showed that the SMCs insulated by Al_2O_3/resin composites have significantly improved resistivity, frequency stability of permeability, core loss, and thermal conductivity. This work provides a viable solution for the fabrication of high magnetic performance and temperature-resistant SMCs.

2. Experimental Methods

2.1. SMC Preparation

The raw FeSiCr powders with a mean particle size of 10 μm were commercially purchased from Antai Technology Co., Ltd (Beijing, China)., and they had the composition of 88.5–90.5 wt.% Fe, 4.5–5.5 wt.% Si, and 5–6 wt.% Cr. $Al(OPr^i)_3$ (Macklin) was used as the precursor for the hydrolysis reaction to obtain the Al_2O_3. Silicone (KR5235) and epoxy resin (AFG90H) supplied by Shin-Etsu Chemical Co., Ltd (Tokyo, Japan). and Tiantai New Materials Co., Ltd (Guangzhou, China)., respectively, were used as the binder to improve the moldability and strength of the powders. KH550 (Usolf) silane coupling agent was used to improve the binding between the organic and inorganic composite coatings. All chemicals used were of an analytical grade and could be used without further purification in this work.

The $Al(OPr^i)_3$, ground into powder, was added to deionized water and heated to 85 °C with mechanical stirring. After 30 min, nitric acid was added as a precipitating solubilizer and catalyst. Afterward, stable and transparent AlOOH sol was obtained by continuous mechanical stirring at 85 °C for 5 h. The molar ratio of $Al(OPr^i)_3$, deionized water, and nitric acid was 1:90:0.3. AlOOH sols with different solid contents (0.5–2.0 wt.%) were mixed and coated with FeSiCr powders under 105 °C until the solvent was evaporated. The mixed powders were subsequently calcined at 700 °C under a high vacuum pure argon atmosphere for 1 h to obtain the Al_2O_3 layer. Phosphated and SiO_2-coated samples with similar resin bonding were fabricated by the conventional phosphating process [26] and TEOS hydrolysis [5], respectively, to compare the thermal conductivity.

The organic coating solution was prepared by dissolving 1 wt.% epoxy-modified silicone resin and 0.5 wt.% KH550 in acetone. The previously obtained powders were dipped into the organic coating solution and stirred until the acetone evaporated. Subsequently, the composite powders and 0.5 wt.% of the lubricant barium stearate were compacted at 1200 MPa into toroidal-shaped SMCs with the dimensions of Φ20 × Φ12 × 4.6 mm. It should be mentioned that the size of our tested samples was not based on the international standards but on the technique requirements of Mentech Optical & Magnetic Co., Ltd (Dongguan, China). Similar sample dimensions have also been used in some studies of FeSiCr SMCs from other research groups [12,13,15]. Finally, the SMCs were heated to 180 °C for 2 h to cure the resin. A schematic diagram of the preparation process of FeSiCr SMCs is presented in Figure 1.

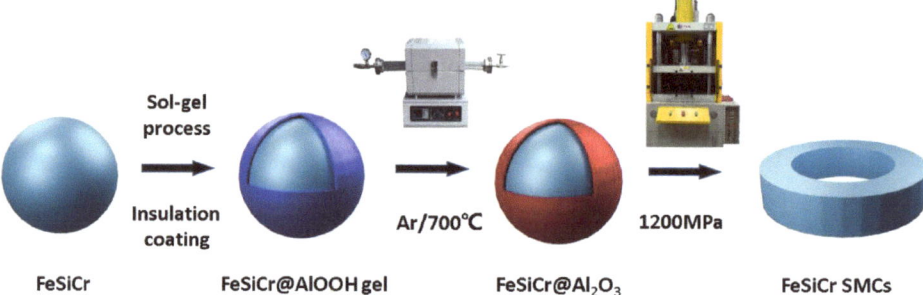

Figure 1. Schematic diagram of the preparation process of FeSiCr SMCs.

2.2. Characterizations

The surface morphologies of the powders and a cross-section of the SMCs were observed by scanning electron microscopy (SEM, FEI Quanta 200, Hillsboro, OR, USA) with energy dispersive X-ray spectroscopy (EDS, EDAX Genesis Xm 2, Pleasanton, USA). Differential thermal analysis (DSC, STA449C, Netzsch, Bavaria, Germany) was used to scan the thermal variation of the FeSiCr coated by AlOOH gel in an argon atmosphere heated to 900 °C at a heating rate of 10 K/min. Fourier transform infrared spectroscopy (FTIR, VERTEX 70, Bruker, Billerica, MA, USA) was used to investigate the chemical structure and state of the coatings at 400–4000 cm^{-1} transmittance. The phase structures of powders before and after insulation treatment were confirmed by X-ray diffraction (XRD, Philips X' Pert, PANalytical, Almelo, The Netherlands) using Cu Kα radiation at room temperature. The electrical resistivity was measured by a high resistance weak current tester (ST2643, Jingge, Suzhou, China). The density was calculated based on the weight and size of the SMC core. The hysteresis loops were measured at room temperature using a Vibrating Sample Magnetometer (VSM-3105, East Changing, Beijing, China). An impedance analyzer (Agilent E4990A, Keysight, Beijing, China) was used to measure the permeability μ_e and quality factor Q of SMCs from 20 kHz to 10 MHz, with contact electrodes in a double-ended configuration. μ_e can be defined as follows:

$$\mu_e = \frac{L}{2h \times \ln(D/d) \times N^2 \times 10^{-10}} \quad (1)$$

where L is the effective self-inductance, N is the number of copper wire turns, h is the height of the sample, and D and d are the outer and inner diameters of the sample, respectively. The core losses were measured using a soft magnetic AC test set (MATS-3010SA, Linkjoin, Loudi, China) at B_m = 50 mT, 20–200 kHz. The thermal conductivity of strip block-pressed SMCs was measured using a physical property measurement system (PPMS-9).

3. Results and Discussion

3.1. Structure Analysis

By the hydrolysis of Al(OPr^i)$_3$, the Al$_2$O$_3$ insulating layer was coated on the FeSiCr powders [27]. The main hydrolysis process can be described with the following three equations:

$$Al(C_3H_7O)_3 + H_2O \rightarrow Al(C_3H_7O)_2(OH) + C_3H_7OH \tag{2}$$

$$2Al(C_3H_7O)_2(OH) + H_2O \rightarrow Al_2O(C_3H_7O)_2OH_2 + 2C_3H_7OH \tag{3}$$

$$Al_2O(C_3H_7O)_2OH_2 + H_2O \rightarrow 2AlOOH + 2C_3H_7OH \tag{4}$$

The total reaction is:

$$Al(C_3H_7O)_3 + 2H_2O \rightarrow AlOOH + 3C_3H_7OH \tag{5}$$

Since the whole hydrolysis reaction is conducted in solution, the resulting product of boehmite (AlOOH) particles is easily and uniformly dispersed in the solution at the molecular level. During preparation, these particles cover the powders as an insulating layer, which will be pyrolyzed into Al$_2$O$_3$ by subsequent annealing. The by-product of propyl alcohol has a low boiling point and thus mostly evaporates during the heating.

During the sol-gel process, the solute is aged, and the colloidal pellets slowly polymerize to form a gel with a three-dimensional spatial network structure. After drying, sintering, and curing the gel, a molecular and even nanostructured coating layer could be formed on the powders. To explore the dehydration temperature of the AlOOH gel, the DSC curve of the FeSiCr powders coated by AlOOH gel was tested, as shown in Figure 2. Although the insulation layer is thin, there is still an obvious heat absorption peak with an area of ΔH = 1.47 J/g of heat absorption per unit, which is obtained by calculating the peak area on the baseline based on the integration. The decomposition of boehmite gel begins at 660 °C, and this reaction is practically complete by 707 °C. This reaction can be described by the following equation [28]:

$$2\alpha - AlOOH \rightarrow \gamma - Al_2O_3 + H_2O \tag{6}$$

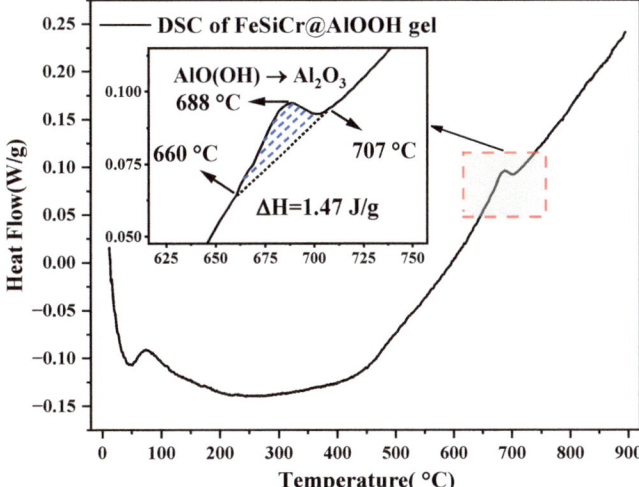

Figure 2. The DSC curve of the FeSiCr powder coated by AlOOH gel.

A lower annealing temperature is needed not only for the consideration of energy savings but also to minimize partial oxidation, which may deteriorate the magnetic permeability. The peak point of the DSC curve corresponds to the fastest energy change, which is 688 °C, as shown in Figure 2. In this work, a temperature of 700 °C, slightly higher than 688 °C, was chosen as the heating temperature to obtain more physically stable Al_2O_3 coating.

FTIR analysis was performed on the FeSiCr powders coated by AlOOH gel before and after heating at 700 °C to understand the evolution of the insulating coatings during the heating process, as shown in Figure 3. The FTIR spectrum, taken from the uncoated powders, is also shown in Figure 3 for comparison. The broad absorption bands around 3430 cm^{-1} and 1630 cm^{-1} are attributed to the stretching of -OH [29], while the weakening of the -OH vibration of the coated powders is due to the formation of Al-O-Al metal bonds by dehydration of the solute with the surface -OH sites. The absorption bands at 490 cm^{-1} and 435 cm^{-1} are attributed to the Si-O and Fe-O vibrations due to trace oxidation appearing on all sample surfaces [5,30]. Compared to uncoated powder, the new 1060 cm^{-1} and 1160 cm^{-1} absorption bands correspond to the Al-OH vibration of the boehmite AlOOH [19,31], and the absorption band around 609 cm^{-1} corresponds to Al-O-Al [32,33]. It indicates that an insulating layer of AlOOH gel and Al_2O_3 has been formed on the powder surface [31,33]. Due to the high-temperature pyrolysis that converts the AlOOH gel into alumina, the sample after heating at 700 °C for 1 h showed a significantly enhanced Al-O-Al vibration at 609 cm^{-1} compared to the untreated sample. At the same time, the bands around 1380 cm^{-1} and 887 cm^{-1} originating from Fe-OH and Fe-O were observed due to a small amount of oxidation of the powder after heating [34].

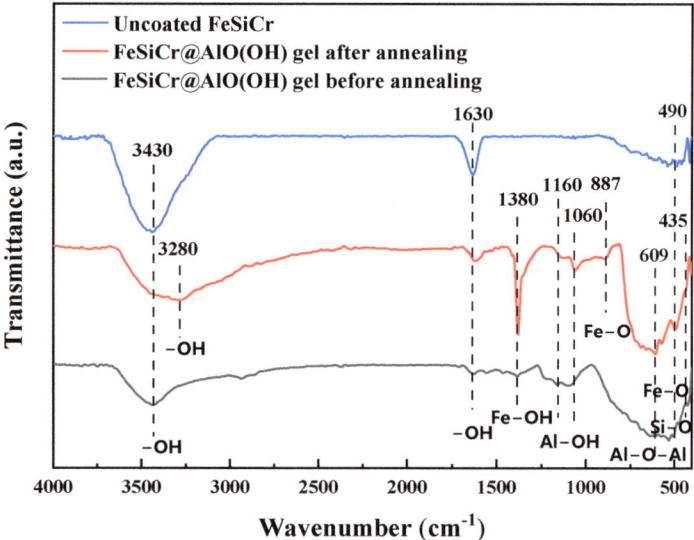

Figure 3. FTIR spectra of the raw FeSiCr powders and the powders insulated by AlOOH gel before and after annealing at 700 °C for 1 h.

The phase structures of powders before and after insulation treatment are shown in Figure 4. The uncoated FeSiCr powder and all Al_2O_3 coated composite powders exhibited sharp crystallization peaks at 44.8°, 65.2°, and 82.6° with crystallographic indices of (110), (200), and (211), respectively, which are typical of the characteristic peaks of α-Fe(Si,Cr). The obvious Al_2O_3 diffraction peaks were not clearly detected. The reasons can be attributed to the content of the formed Al_2O_3 being very low or the heating temperature not reaching the crystallization temperature of Al_2O_3. Nevertheless, this outcome does not disprove that there is an alumina insulation layer generated on the powder surface.

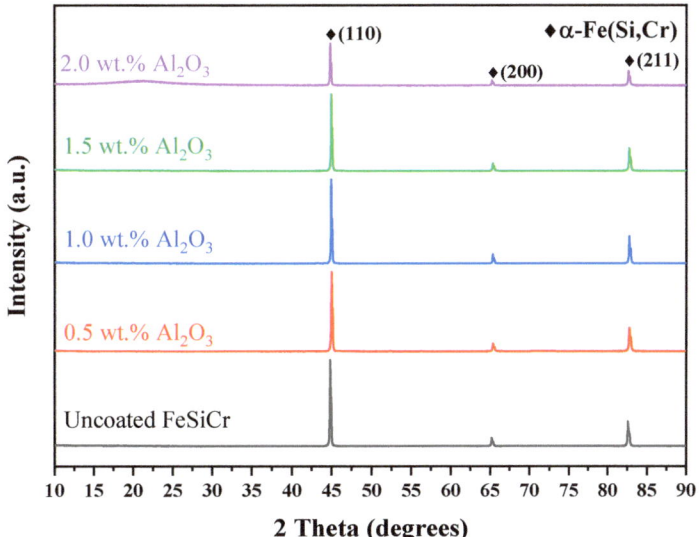

Figure 4. X-ray diffraction (XRD) of uncoated and different Al_2O_3-coated FeSiCr powders.

Figure 5 shows the morphologies and EDS elemental distribution of the FeSiCr powders insulated by different concentrations of the Al_2O_3 layer. The 0.5 wt.% Al_2O_3-coated powder showed a slight, rough surface, and no significant change could be observed compared to the raw powders in Figure 5a because the thickness of the layer is very thin. In Figure 5b, the detected Fe, Si, and Cr elements came from the initial powder. The Al and O elements exhibited the same uniform distribution. The mass ratio of the Al element increased with the gradual increase in the solid content of Al_2O_3 in the boehmite sols. At the same time, a thicker surface insulation layer is clearly revealed in Figure 5c,d. The 1.5 wt.% Al_2O_3-coated powder became rough, and a few visible flocs gathering appeared on the powder surface, as shown in Figure 5d. The mass ratio of the Al element is significantly increased compared to the low concentration of the Al_2O_3-coated powder. According to the SEM results, there is a thin Al_2O_3 insulation layer covering the powder, and the thickness was affected by the Al_2O_3 concentration. To provide maximum permeability and density, the amount of interparticle insulation should be minimized. A thin and uniform insulation layer is generally required to minimize the eddy current in high-frequency applications.

The polished cross-section of the SMCs is shown in Figure 6. The EDS result of line scanning crossing the powder interface indicates that the Al and O elements appear at the interface of particles containing Fe elements. The elemental distribution on particle boundaries indicates that the particles are isolated by the alumina insulating layer. The coating layer forms the isolation network between ferrous particles, which can reduce the powder contact and even act as a heat conduction channel. In addition, air gaps can still be found between the particles under pressure of 1200 MPa. Partial extrusion deformation can also be observed. Partial deformation can reduce the porosity and increase the density, but this additional deformation induces high stress and thus may increase the hysteresis loss [35]. The internal stress introduced during compaction can often be relaxed in the post-annealing process.

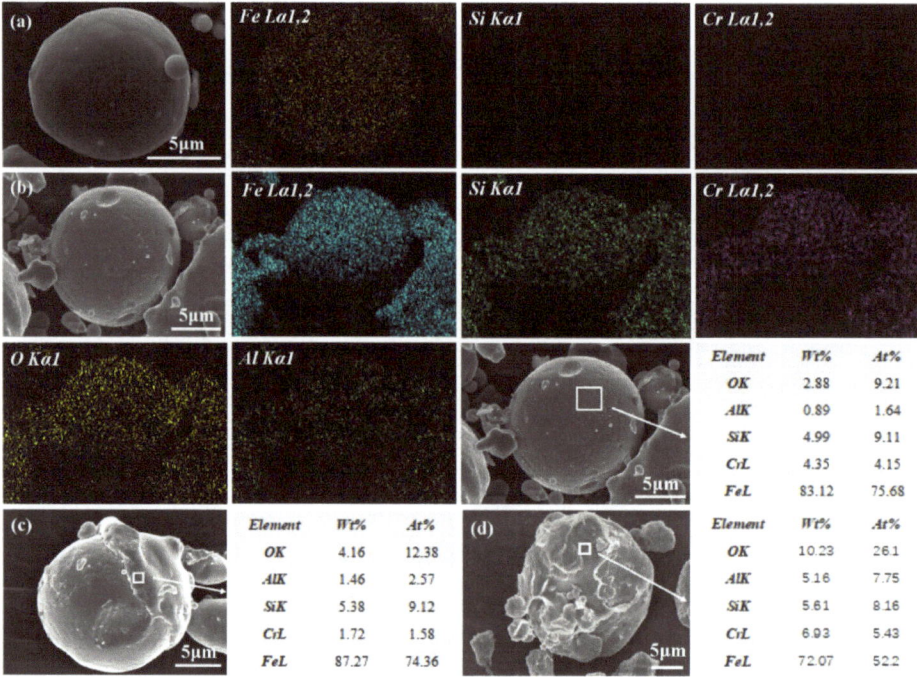

Figure 5. SEM micrograph and EDS elemental distribution maps of the FeSiCr powders insulated by different Al$_2$O$_3$ concentrations: (**a**) raw powder; (**b**) 0.5%; (**c**) 1.0%; and (**d**) 1.5%.

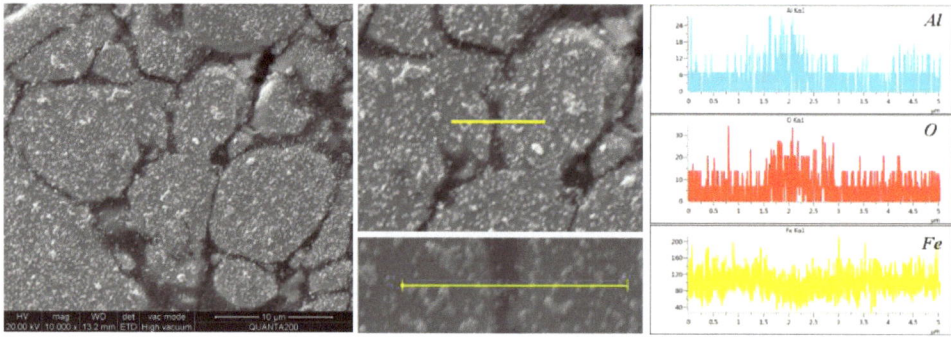

Figure 6. SEM micrograph and EDS analysis (line scanning) of the cross-section of FeSiCr@Al$_2$O$_3$ SMCs.

3.2. Magnetic Performance

Figure 7 illustrates the magnetic hysteresis loops of FeSiCr powders coated with different Al$_2$O$_3$ contents. The enlarged image shows the saturation magnetization M_s. Compared to the uncoated powders, as the content of Al$_2$O$_3$ increases to 2.0 wt.%, the M_s value decreases monotonically from 193 emu/g to 177 emu/g due to the magnetic dilution effect of the non-magnetic substance Al$_2$O$_3$. This result indicates that the introduction of inorganic insulation coating does not significantly reduce the M_s of the magnetic powder, and M_s remains in the normal level of FeSiCr powder (130–200 emu/g) [12]. Combined with the content and morphology of the insulating layer within Figure 5, the introduced Al$_2$O$_3$ insulating layer is thin, with less influence on the deterioration of the original powder M_s.

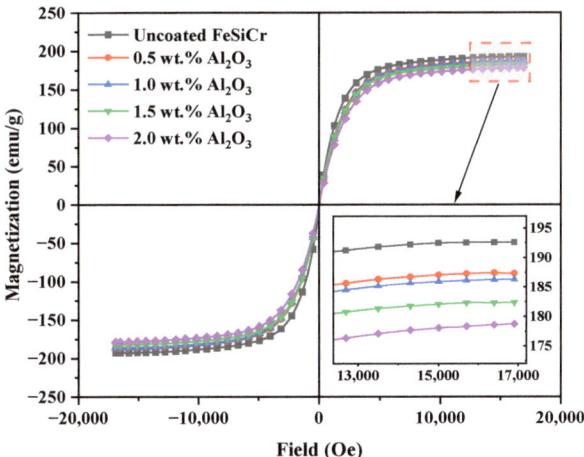

Figure 7. The magnetic hysteresis loops for different Al_2O_3-coated FeSiCr powders.

The effective permeability μ_e variation with the frequency for different contents of Al_2O_3 coatings are shown in Figure 8. The samples fabricated by raw powders exhibit poor frequency stability of μ_e, and as the frequency increases to 5 MHz, the μ_e decrease monotonically drops from 55 to 21 (a decrease of 62%) subjected to the deteriorating effect of eddy currents. All the Al_2O_3/resin-coated SMCs have almost constant permeability from 20 kHz to 5 MHz, due to the good insulation between the magnetic powders. After a 1 wt.% resin coating, the μ_e decreases from 55 to 40 (a decrease of 28%), while a further reduction from 40 to 34 (a decrease of 15%) appears with the increasing concentration (0–2 wt.%) of Al_2O_3. This decrease is due to the magnetic dilution effect of Al_2O_3 and resin, as well as the introduction of more air gaps and cracks into the insulation layer. It can be concluded from the reduction percentage of μ_e that the addition of resin has a greater effect on the decrease in magnetic permeability than the Al_2O_3 layer. Thus, the 0.5 wt.% Al_2O_3/resin coating sample causes a minor decrease in μ_e compared to the resin-only coating sample, whereas the μ_e decreased only from 40 to 39.

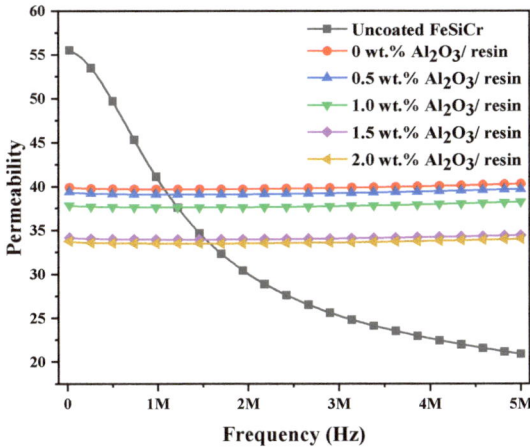

Figure 8. Permeability of FeSiCr SMCs with different contents of Al_2O_3 insulation versus frequency (from 20 kHz to 5 MHz).

The core loss is the dissipating part of energy that is converted irreversibly into heat through the periodically magnetized process. In the current application frequency field, the total loss mainly consists of hysteresis loss, eddy current loss, and anomalous loss. Hysteresis loss is due to the irreversible domain wall displacement and magnetization vector rotation. The AC magnetic process generates eddy currents, which depend on the inductive magnetic field and frequency in working conditions. Anomalous loss is a combination of relaxation and resonant losses. According to the classical loss theory, the total loss (W_t) is mainly determined by hysteresis loss (W_h), eddy current loss (W_e), and anomalous loss (W_a) according to the following formula [36,37]:

$$W_t = W_h + W_e + W_a \approx \oint H dB + \frac{C_e B^2 d^2}{\rho} f + C_a f^{1/2} \qquad (7)$$

where f is the frequency, H is the magnetic field strength, B is the magnetic induction, C_e and C_a are the proportionality constants, ρ is the resistivity, and d is the thickness of the material. Figure 9 shows the variation of core loss versus frequencies from 20 to 200 kHz at B_m = 50 mT for FeSiCr SMCs with 0–2 wt.% Al_2O_3/resin coatings. The anomalous loss calculated by fitting the experimental results is very small and neglected, which is only important at very low induction levels and very high frequencies [2]. The raw powder pressed sample exhibits the highest total loss, which severely worsens to 4050 kW/m³ with the increase in frequency to 200 kHz. The presence of a large, nonlinear eddy current loss fraction indicates a large heat generation effect of inter-particle eddy currents without insulation treatment. Compared to the 1 wt.% resin only-coated SMC sample, the SMCs with (0.5–2 wt.%) Al_2O_3/resin showed a significant reduction in core loss. The minimum value of 1173 mW/cm³ corresponds to the sample with 0.5 wt.% Al_2O_3/resin, which exhibited a 27% reduction compared to the resin only-coated sample and a 71% reduction compared to an uncoated powder sample. This reduction in loss is due to the suppression of eddy currents by the insulation layer. With the increase in Al_2O_3 concentration, the core loss grows gradually because of the increase in hysteresis loss in part. In a ferromagnetic pressing material, Al_2O_3 impurities between the particles and stressed regions give rise to pinning sites that can hinder domain wall motion [38,39], increasing the coercive force and directly increasing the hysteresis loss.

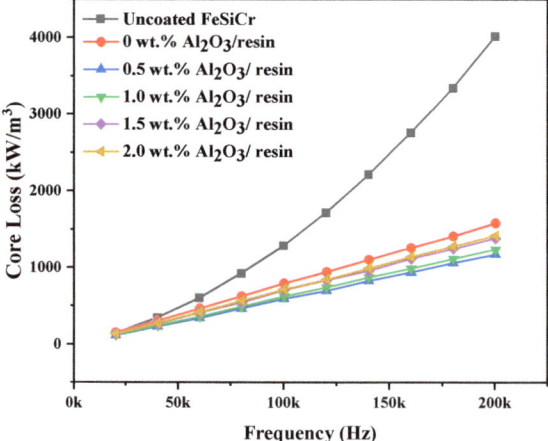

Figure 9. Core losses of FeSiCr SMCs with different contents of Al_2O_3 insulation in the frequency range from 20 kHz to 200 kHz.

Table 1 lists the density, electric resistivity ρ, effective permeability μ_e, and core losses of FeSiCr SMCs insulated with different contents of Al_2O_3. The standard deviations are calculated by five samples from the same group. The density of SMCs monotonically decreases from 6.31 g/cm^3 of the raw powder to 5.97 g/cm^3 with 2 wt.% Al$_2$O$_3$/resin. The reduction is due to the relatively lower-density insulation layer. The resistivity of the samples after resin-only and 2 wt.% Al$_2$O$_3$/resin coatings increases by nearly 4700 and 1,300,000 times, respectively, compared with the uncoated sample, indicating an increasing trend with the Al$_2$O$_3$ addition. The increasing resistivity can reduce the eddy current between the particles, decreasing the heat dissipation and ensuring the soft magnetic performance of SMCs at high frequencies. In fact, for a test frequency not higher than 200 kHz, an appropriately increased resistivity is able to limit the eddy current loss to a low value, but excessive insulation may increase the hysteresis loss and the total losses. Specifically, as the insulation layer content increases, some properties are optimized with sacrifice of other properties. As shown in Table 1, the resistivity and core losses can be improved, but the permeability and saturation magnetization show some reduction as the Al$_2$O$_3$ content increases. The sample with 0.5% Al$_2$O$_3$-resin showed relatively good properties.

Table 1. The density, resistivity ρ, effective permeability μ_e, and core losses of FeSiCr SMCs insulated with different contents of Al$_2$O$_3$.

Sample (Powder Treatment)	Density (g/cm^3)	Resistivity ρ ($\Omega \cdot$cm)	Permeability (μ_e/200 kHz)	Ps@50 mT/mW/cm^3	
				100 kHz	200 kHz
Uncoated FeSiCr	6.31 ± 0.03	58 ± 9	55.5 ± 1.6	1283 ± 8	4015 ± 26
0% Al$_2$O$_3$-resin	6.23 ± 0.03	(2.8 ± 0.1) × 10^5	39.9 ± 1.2	783 ± 3	1584 ± 11
0.5% Al$_2$O$_3$-resin	6.19 ± 0.02	(8.3 ± 0.5) × 10^5	39.3 ± 1.3	589 ± 3	1173 ± 8
1.0% Al$_2$O$_3$-resin	6.10 ± 0.02	(2.1 ± 0.2) × 10^6	37.8 ± 1.1	620 ± 5	1239 ± 10
1.5% Al$_2$O$_3$-resin	5.98 ± 0.02	(1.9 ± 0.1) × 10^7	34.2 ± 0.9	701 ± 2	1385 ± 5
2.0% Al$_2$O$_3$-resin	5.97 ± 0.02	(7.8 ± 0.2) × 10^7	33.7 ± 0.9	712 ± 5	1420 ± 9

With the demand for high energy storage and low energy dissipation at high frequencies in industrialization, the conversion efficiency is an important characteristic of SMCs. Quality factor Q is a vital parameter for the electric component in the circuit. It represents the ratio of energy storage to energy dissipation in inductor devices: a higher quality factor implies better high-frequency soft magnetic performance, which can be understood as follows [40]:

$$Q = \frac{\mu'}{\mu''} \quad (8)$$

where μ' and μ'' are the real and imaginary parts of complex permeability, respectively. Figure 10 shows that the uncoated powder reaches the peak Q = 10.3 point at a low frequency (80 kHz) and continuously decreases as the frequency increases. The sample without insulation treatment is only suitable for power conversion at lower frequencies with a lower efficiency. In contrast, the Q value of the composite insulated SMCs maintains a high level up to 2~3 MHz, indicating good performance for high-frequency applications. The variation of the Q value with Al$_2$O$_3$ content shows the same trend as the core loss (Figure 9), indicating the important, positive effect of insulation treatment for SMCs. The highest quality factor of 94 at 1 MHz also indicates that the 0.5 wt.% Al$_2$O$_3$/resin coating has the best insulation effect.

3.3. Thermal Conductivity

Soft magnetic devices are increasingly required to be used under high electrical current. The thermal stresses due to joule heating and magnetic losses can be damaging. A higher thermal conductivity might improve the thermal performance of the device by increasing the ability to dissipate heat. From the Maxwell–Eucken mixture equation, an effective thermal conductivity of the composite core material was determined by the following parameters [41]:

$$k_e = k_l \cdot \left[\frac{k_p + 2 \cdot k_l + 2 \cdot V_f \cdot (k_p - k_l)}{k_p + 2 \cdot k_l - V_f \cdot (k_p - k_l)} \right] \quad (9)$$

where k_p and k_l represent the thermal conductivity of the iron powder and layer, whereas k_e and V_f represent the effective thermal conductivity and volume fraction of the filling, respectively. It can be concluded that k_e is mainly dependent on the layer material. The choice of a higher thermal conductivity coating layer can effectively improve its k_e.

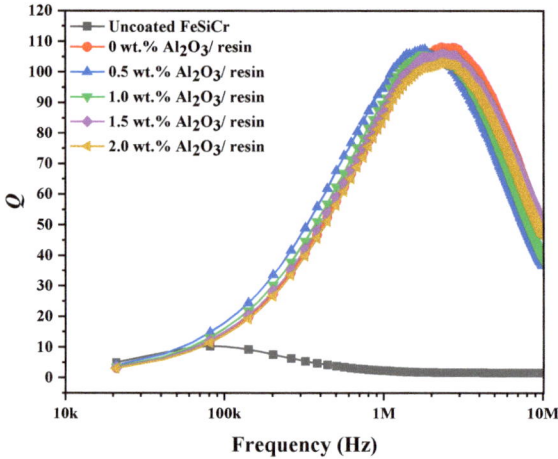

Figure 10. Quality factor Q of FeSiCr SMCs with different contents of Al_2O_3 insulation at the frequency range of 20 kHz to 10 MHz.

The thermal conductivities of FeSiCr-based SMCs with different inorganic layers and similar resin bindings are shown in Figure 11, and they were tested at near room temperature. The thermal conductivity showed a slow decrease with increasing temperature. The conduction of heat within a metal material is mainly achieved by collisions between free electrons. When the temperature increases, the speed of the thermal motion of electrons accelerates, and collisions with lattice dots are more frequent. As a result, the average free range is shortened, and the thermal conductivity decreases. The Al_2O_3-coated sample maintained a higher thermal conductivity, which was ~5.3 W/m·K at room temperature. Using the Maxwell mixture relationship to estimate the thermal conductivity of composites layer k_l, the two parameters were fixed at: $V_f = 0.637$, the maximum fraction to which a volume can be filled with randomly packed spheres; and $k_p = 74$ W/m·K, the thermal conductivity of FeSiCr powders. The calculated thermal conductivity of the Al_2O_3/resin composites layer k_l is 0.92 W/m·K, which is much higher than the thermal conductivity of general resins (0.11–0.53 W/m·K) [25]. As for the two types of inorganic insulation most commonly used in industrial manufacturing, the phosphating process exhibited higher thermal conductivity than the SiO_2 coating method. This finding is consistent with the trends of the thermal conductivity of these two inorganic insulations themselves. Al_2O_3 insulation layer exhibits the best thermal conducting for the SMCs. It is believed that, as an easily prepared metal oxide with high thermal conductivity, Al_2O_3 insulation treatment will have promising applications for SMC inductor production.

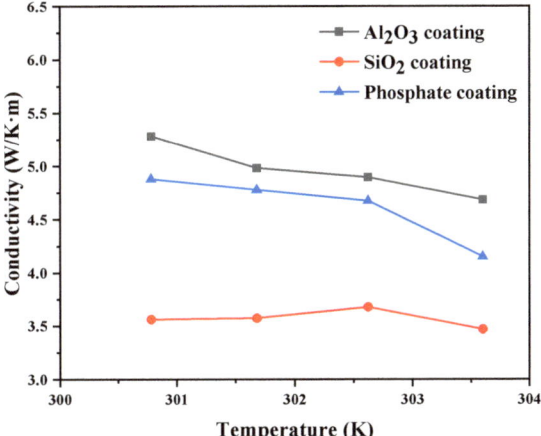

Figure 11. The thermal conductivity of FeSiCr SMCs with different inorganic layers and similar resin bonding values at various temperatures near room temperature.

4. Conclusions

FeSiCr soft magnetic composites (SMCs) based on powder coated with an Al_2O_3/resin insulation layer by the sol-gel method were fabricated. Al_2O_3/resin insulation treatment not only effectively optimized the core loss, resistivity, and quality factor Q of SMCs, but it also improved their thermal conductivity. The sample with 0.5 wt.% Al_2O_3/resin exhibited the optimum comprehensive properties with saturation magnetization M_s = 188 emu/g, effective permeability μ_e = 39, resistivity $\rho = 8.28 \times 10^5$ $\Omega \cdot$cm, quality factor Q = 94 (at 1 MHz), and core losses = 1173 kW/m^3 (at 200 kHz, 50 mT). The thermal conductivity of SMCs is ~5.3 W/m·K at room temperature, which is higher than that of conventional inorganic coating layers, such as phosphate and SiO_2. The use of inorganic insulating layer techniques allows for stability of μ_e and loss reduction at high frequencies, but it inevitably leads to reductions in μ_e and M_s. The low core loss and high stability of μ_e and high thermal conductivity of SMCs constitute considerable improvements for SMC-based devices to be used in high-frequency and high-temperature environments. This work provides a new strategy for developing SMCs that can work at relatively high temperatures with high reliability, which could be beneficial to high frequencies, miniaturization, and integration of electronic components.

Author Contributions: Formal analysis, Q.X. and H.Y. (Hongya Yu); investigation, X.C.; methodology, Q.X., H.Y. (Han Yuan), G.H. and Z.L.; project administration, H.Y. (Hongya Yu) and Z.L.; supervision, H.Y. (Hongya Yu); visualization, Q.X.; writing—original draft, Q.X.; writing—review and editing, Q.X., H.Y. (Hongya Yu), G.H. and Z.L. All authors have read and agreed to the published version of the manuscript.

Funding: This work was supported by the Dongguan Innovative Research Team Program (Grant No. 2020607231010) and the Guangdong Provincial Natural Science Foundation of China (No. 2021A1515010642).

Institutional Review Board Statement: Not applicable.

Informed Consent Statement: Not applicable.

Data Availability Statement: Not applicable.

Conflicts of Interest: The authors declare no conflict of interest.

References

1. Silveyra, J.M.; Ferrara, E.; Huber, D.L.; Monson, T.C. Soft magnetic materials for a sustainable and electrified world. *Science* **2018**, *362*, eaao0195. [CrossRef] [PubMed]
2. Shokrollahi, H.; Janghorban, K. Soft magnetic composite materials (SMCs). *J. Mater. Process. Technol.* **2007**, *189*, 1–12. [CrossRef]
3. Sunday, K.J.; Taheri, M.L. Soft magnetic composites: Recent advancements in the technology. *Met. Powder Rep.* **2017**, *72*, 425–429. [CrossRef]
4. Périgo, E.A.; Weidenfeller, B.; Kollár, P.; Füzer, J. Past, present, and future of soft magnetic composites. *Appl. Phys. Rev.* **2018**, *5*, 31301. [CrossRef]
5. Guan, W.W.; Shi, X.Y.; Xu, T.T.; Wan, K.; Zhang, B.W.; Liu, W.; Su, H.L.; Zou, Z.Q.; Du, Y.W. Synthesis of well-insulated Fe–Si–Al soft magnetic composites via a silane-assisted organic/inorganic composite coating route. *J. Phys. Chem. Solids* **2021**, *150*, 109841. [CrossRef]
6. Li, W.; Cai, H.; Kang, Y.; Ying, Y.; Yu, J.; Zheng, J.; Qiao, L.; Jiang, Y.; Che, S. High permeability and low loss bioinspired soft magnetic composites with nacre-like structure for high frequency applications. *Acta Mater.* **2019**, *167*, 267–274. [CrossRef]
7. Li, W.; Li, W.; Ying, Y.; Yu, J.; Zheng, J.; Qiao, L.; Li, J.; Zhang, L.; Fan, L.; Wakiya, N.; et al. Magnetic and Mechanical Properties of Iron-Based Soft Magnetic Composites Coated with Silane Synergized by Bi_2O_3. *J. Electron. Mater.* **2021**, *50*, 2425–2435. [CrossRef]
8. Yao, Z.; Peng, Y.; Xia, C.; Yi, X.; Mao, S.; Zhang, M. The effect of calcination temperature on microstructure and properties of FeNiMo@Al_2O_3 soft magnetic composites prepared by sol-gel method. *J. Alloy. Compd.* **2020**, *827*, 154345. [CrossRef]
9. Li, J.; Ni, J.; Zhu, S.; Feng, S. Evolution of magnetic loss with annealing temperature in FeSiAl/carbonyl iron soft magnetic composite. *Mater. Technol.* **2022**, *37*, 2313–2317. [CrossRef]
10. Guo, R.; Wang, S.; Yu, Z.; Sun, K.; Jiang, X.; Wu, G.; Wu, C.; Lan, Z. FeSiCr@NiZn SMCs with ultra-low core losses, high resistivity for high frequency applications. *J. Alloy. Compd.* **2020**, *830*, 154736. [CrossRef]
11. Nie, W.; Yu, T.; Wang, Z.; Wei, X. High-performance core-shell-type FeSiCr@MnZn soft magnetic composites for high-frequency applications. *J. Alloy. Compd.* **2021**, *864*, 158215. [CrossRef]
12. Wang, F.; Dong, Y.; Chang, L.; Pan, Y.; Chi, Q.; Gong, M.; Li, J.; He, A.; Wang, X. High performance of Fe-based soft magnetic composites coated with novel nano-$CaCO_3$/epoxy nanocomposites insulating layer. *J. Solid State Chem.* **2021**, *304*, 122634. [CrossRef]
13. Dong, B.; Qin, W.; Su, Y.; Wang, X. Magnetic properties of FeSiCr@MgO soft magnetic composites prepared by magnesium acetate pyrolysis for high-frequency applications. *J. Magn. Magn. Mater.* **2021**, *539*, 168350. [CrossRef]
14. Hsiang, H.-I.; Chuang, K.-H.; Lee, W.-H. FeSiCr Alloy Powder to Carbonyl Iron Powder Mixing Ratio Effects on the Magnetic Properties of the Iron-Based Alloy Powder Cores Prepared Using Screen Printing. *Materials* **2021**, *14*, 1034. [CrossRef]
15. Gong, M.; Dong, Y.; Huang, J.; Chang, L.; Pan, Y.; Wang, F.; He, A.; Li, J.; Liu, X.; Wang, X. The enhanced magnetic properties of FeSiCr powder cores composited with carbonyl iron powder. *J. Mater. Sci. Mater. Electron.* **2021**, *32*, 8829–8836. [CrossRef]
16. Yu, H.; Zhou, S.; Zhang, G.; Dong, B.; Meng, L.; Li, Z.; Dong, Y.; Cao, X. The phosphating effect on the properties of FeSiCr alloy powder. *J. Magn. Magn. Mater.* **2022**, *552*, 168741. [CrossRef]
17. Xia, C.; Peng, Y.; Yi, Y.; Deng, H.; Zhu, Y.; Hu, G. The magnetic properties and microstructure of phosphated amorphous FeSiCr/silane soft magnetic composite. *J. Magn. Magn. Mater.* **2019**, *474*, 424–433. [CrossRef]
18. Yaghtin, M.; Taghvaei, A.H.; Hashemi, B.; Janghorban, K. Structural and magnetic properties of Fe-Al2O3 soft magnetic composites prepared by sol-gel method. *Int. J. Mater. Res.* **2013**, *105*, 474–479. [CrossRef]
19. Liu, D.; Wu, C.; Yan, M.; Wang, J. Correlating the microstructure, growth mechanism and magnetic properties of FeSiAl soft magnetic composites fabricated via HNO_3 oxidation. *Acta Mater.* **2018**, *146*, 294–303. [CrossRef]
20. Monier-Vinard, E.; Bissuel, V.; Laraqi, N.; Daniel, O.; Signing, D. Experimental characterization of DELPHI Compact Thermal Model for Surface-Mounted Soft Magnetic Composite Inductor. In Proceedings of the 2015 21st International Workshop on Thermal Investigations of ICs and Systems (THERMINIC), Paris, France, 30 September–2 October 2015.
21. Ouyang, Y.; Hou, G.; Bai, L.; Li, B.; Yuan, F. Constructing continuous networks by branched alumina for enhanced thermal conductivity of polymer composites. *Compos. Sci. Technol.* **2018**, *165*, 307–313. [CrossRef]
22. Yu, L.; Lirui, S.; Qingyu, W.; Haitao, Q.; Chuncheng, H.; Qingquan, L. Effect of branched alumina on thermal conductivity of epoxy resin. *J. Ind. Eng. Chem.* **2022**, *120*, 209–215.
23. Pinheiro, A.; da Costa, Z.; Bell, M.; Anjos, V.; Dantas, N.; Reis, S. Thermal characterization of iron phosphate glasses for nuclear waste disposal. *Opt. Mater.* **2011**, *33*, 1975–1979. [CrossRef]
24. Chen, H.; Ginzburg, V.V.; Yang, J.; Yang, Y.; Liu, W.; Huang, Y.; Du, L.; Chen, B. Thermal conductivity of polymer-based composites: Fundamentals and applications. *Prog. Polym. Sci.* **2016**, *59*, 41–85. [CrossRef]
25. Joshi, G.; Saxena, N.; Mangal, R. Temperature dependence of effective thermal conductivity and effective thermal diffusivity of Ni-Zn ferrites. *Acta Mater.* **2003**, *51*, 2569–2576. [CrossRef]
26. Li, K.; Cheng, D.; Yu, H.; Liu, Z. Process optimization and magnetic properties of soft magnetic composite cores based on phosphated and mixed resin coated Fe powders. *J. Magn. Magn. Mater.* **2020**, *501*, 166455. [CrossRef]
27. Hu, B.; Jia, E.; Du, B.; Yin, Y. A new sol-gel route to prepare dense Al_2O_3 thin films. *Ceram. Int.* **2016**, *42*, 16867–16871. [CrossRef]
28. Paulik, F.; Paulik, J.; Naumann, R.; Köhnke, K.; Petzold, D. Mechanism and kinetics of the dehydration of hydrargillites. Part I. *Thermochim. Acta* **1983**, *64*, 1–14. [CrossRef]

29. Padmaja, P.; Anilkumar, G.; Mukundan, P.; Aruldhas, G.; Warrier, K. Characterisation of stoichiometric sol–gel mullite by fourier transform infrared spectroscopy. *Int. J. Inorg. Mater.* **2001**, *3*, 693–698. [CrossRef]
30. Pu, H.; Jiang, F.; Yang, Z. Studies on preparation and chemical stability of reduced iron particles encapsulated with polysiloxane nano-films. *Mater. Lett.* **2006**, *60*, 94–97. [CrossRef]
31. Shen, S.-C.; Ng, W.K.; Zhong, Z.-Y.; Dong, Y.-C.; Chia, L.; Tan, R.B.H. Solid-Based Hydrothermal Synthesis and Characterization of Alumina Nanofibers with Controllable Aspect Ratios. *J. Am. Ceram. Soc.* **2009**, *92*, 1311–1316. [CrossRef]
32. Peng, Y.; Yi, Y.; Li, L.; Yi, J.; Nie, J.; Bao, C. Iron-based soft magnetic composites with Al_2O_3 insulation coating produced using sol–gel method. *Mater. Des.* **2016**, *109*, 390–395. [CrossRef]
33. Liu, D.; Wu, C.; Yan, M. Investigation on sol–gel Al2O3 and hybrid phosphate-alumina insulation coatings for FeSiAl soft magnetic composites. *J. Mater. Sci.* **2015**, *50*, 6559–6566. [CrossRef]
34. Lefèvre, G. In situ Fourier-transform infrared spectroscopy studies of inorganic ions adsorption on metal oxides and hydroxides. *Adv. Colloid Interface Sci.* **2004**, *107*, 109–123. [CrossRef] [PubMed]
35. Sunday, K.J.; Taheri, M.L. NiZnCu-ferrite coated iron powder for soft magnetic composite applications. *J. Magn. Magn. Mater.* **2018**, *463*, 1–6. [CrossRef]
36. Olekšáková, D.; Kollár, P.; Jakubčin, M.; Füzer, J.; Tkáč, M.; Slovenský, P.; Bureš, R.; Fáberová, M. Energy loss separation in NiFeMo compacts with smoothed powders according to Landgraf's and Bertotti's theories. *J. Mater. Sci.* **2021**, *56*, 12835–12844. [CrossRef]
37. Olekšáková, D.; Kollár, P.; Neslušan, M.; Jakubčin, M.; Füzer, J.; Bureš, R.; Fáberová, M. Impact of the Surface Irregularities of NiFeMo Compacted Powder Particles on Irreversible Magnetization Processes. *Materials* **2022**, *15*, 8937. [CrossRef]
38. Kollár, P.; Birčáková, Z.; Füzer, J.; Bureš, R.; Fáberová, M. Power loss separation in Fe-based composite materials. *J. Magn. Magn. Mater.* **2013**, *327*, 146–150. [CrossRef]
39. Talaat, A.; Suraj, M.; Byerly, K.; Wang, A.; Wang, Y.; Lee, J.; Ohodnicki, P. Review on soft magnetic metal and inorganic oxide nanocomposites for power applications. *J. Alloy. Compd.* **2021**, *870*, 159500. [CrossRef]
40. Hsiang, H.-I. Progress in materials and processes of multilayer power inductors. *J. Mater. Sci. Mater. Electron.* **2020**, *31*, 16089–16110. [CrossRef]
41. Monier-Vinard, E.; Bissuel, V.; Dia, C.T.; Daniel, O.; Laraqi, N. Investigation of Delphi Compact Thermal Model Style for Modeling Surface-Mounted Soft Magnetic Composite Inductor. In Proceedings of the 19th International Workshop on Thermal Investigations of ICs and Systems (THERMINIC), Berlin, Germany, 25–27 September 2013.

Disclaimer/Publisher's Note: The statements, opinions and data contained in all publications are solely those of the individual author(s) and contributor(s) and not of MDPI and/or the editor(s). MDPI and/or the editor(s) disclaim responsibility for any injury to people or property resulting from any ideas, methods, instructions or products referred to in the content.

Article

Enhancing the Properties of FeSiBCr Amorphous Soft Magnetic Composites by Annealing Treatments

Hongya Yu [1,2,3,*], Jiaming Li [2], Jingzhou Li [1], Xi Chen [1,2,3], Guangze Han [1,2,3], Jianmin Yang [1] and Rongyin Chen [1]

1. Dongguan Mentech Optical & Magnetic Co., Ltd., Dongguan 523330, China; justin.li@mnc-tek.com.cn (J.L.); xichen@scut.edu.cn (X.C.); phgzhan@scut.edu.cn (G.H.); jimi.yang@mnc-tek.com.cn (J.Y.); chen.rong.yin@mnc-tek.com.cn (R.C.)
2. School of Materials Science and Engineering, South China University of Technology, Guangzhou 510640, China; johnlee1215@163.com
3. South China Institute of Collaborative Innovation, Dongguan 523808, China
* Correspondence: yuhongya@scut.edu.cn

Abstract: Fe-based amorphous powder cores (AMPCs) were prepared from FeSiBCr amorphous powders with phosphate–resin hybrid coating. The high-frequency magnetic properties of AMPCs annealed at different temperatures were systematically studied. After annealing at low temperatures, the effective permeability and core loss improved due to the internal stress of the powder cores being released. The sample annealed at 480 °C exhibits the lowest hysteresis loss of about 29.6 mW/cm^3 at 800 kHz as well as a maximum effective permeability of 36.4, remaining stable until 3 MHz, which could be useful for high-frequency applications.

Keywords: amorphous powder cores; high frequency; internal stress; annealing

Citation: Yu, H.; Li, J.; Li, J.; Chen, X.; Han, G.; Yang, J.; Chen, R. Enhancing the Properties of FeSiBCr Amorphous Soft Magnetic Composites by Annealing Treatments. *Metals* **2022**, *12*, 828. https://doi.org/10.3390/met12050828

Academic Editor: Joan-Josep Suñol

Received: 7 April 2022
Accepted: 4 May 2022
Published: 11 May 2022

Publisher's Note: MDPI stays neutral with regard to jurisdictional claims in published maps and institutional affiliations.

Copyright: © 2022 by the authors. Licensee MDPI, Basel, Switzerland. This article is an open access article distributed under the terms and conditions of the Creative Commons Attribution (CC BY) license (https://creativecommons.org/licenses/by/4.0/).

1. Introduction

Soft magnetic composites (SMCs) are widely used in electronic devices and components in the field of energy conversions, such as transformers, inductors and electrical motors [1–3]. They are key to the efficient operation of the next generation of electrical machines due to their characteristics, such as magnetic and thermal isotropy, high resistivity and high saturation magnetization [2,4]. In order to be used in high-frequency ranges, SMCs require excellent electromagnetic properties, such as high-frequency stability, low core loss and usability with high currents [5]. The frequency characteristics of permeability and total core loss can be significantly affected by structure, density, non-magnetic insulation coatings, internal stress and so on [6]. Generally, insulation coatings can improve the resistivity of powders and insulate the particles, which can reduce the eddy current loss at high frequency. Meanwhile, the non-magnetic insulation coatings will dilute saturation magnetization and decrease permeability. Furthermore, the internal stress generated during the pressing process will hinder domain walls' motion, resulting in the deterioration of coercivity and hysteresis loss [7–9].

There are many types of SMCs, such as ferrites, FeSi, Sendust (FeSiAl) and Fe-based amorphous cores. Compared with traditional soft magnetic alloys, Fe-based amorphous materials have low coercivity, high resistivity and high saturation magnetization [10]. Fe-based amorphous bulks have been used in transformers and are estimated to be able to save approximately 30% of electrical energy. Therefore, Fe-based amorphous materials have attracted widespread attention and are recommended as ideal soft materials for high-frequency applications. However, due to the poor plastic deformation ability of amorphous powders, they require higher pressure during pressing than compared to traditional SMCs. The internal stress generated during the pressing process leads to low permeability and high core loss. In this research, FeSiBCr amorphous magnetic powder cores (AMPCs) with phosphate–resin hybrid insulation coating were fabricated. The annealing effects on the microstructure and magnetic properties of the powder cores were studied systematically.

2. Experimental Procedure

The FeSiBCr gas-atomized amorphous powders with a median particle size of 11 μm were first mixed with phosphoric acid diluted in ethanol with a concentration of 0.6 wt.%. The powders were stirred in the solution at 55 °C for 30 min to obtain the phosphate coating. Then, the coated powders were dried at 120 °C for 30 min. After drying, the coated powders were mixed with 3 wt.% epoxy-modified silicone resin acetone solution to obtain inorganic–organic core–shell composite powders. The composite powders were then compacted into toroidal cores with dimensions of Φ20 mm × Φ12 mm × 5 mm at 1800 MPa. The coated amorphous powder and the powder cores were annealed at 440 °C, 480 °C and 520 °C under the protection of argon for 1 h.

The morphology of the coated powders was characterized by a scanning electron microscope (SEM). The phase identifications for all powders were conducted by an X-ray diffractometer using Cu Kα radiation over the 2θ range of 10°–90°. The saturation magnetizations of all samples were measured with a Physical Property Measurement System (PPMS) equipped with a vibrating sample magnetometer (VSM). The DC hysteresis loops of all powder cores were collected by a soft magnetic direct current test system. The effective permeabilities and quality factors of the powder cores were measured by an impedance analyzer. The core losses of all samples were measured using an AC B-H loop tracer.

3. Results and Discussion

Figure 1 shows SEM images of the raw amorphous powder, the phosphated powder and the phosphated powders after annealing at different temperatures. After phosphating, the surfaces of the powders became rougher than that of the raw powders. However, the morphology of the phosphated powders annealed below 480 °C did not reflect significant changes. In particular, it can be seen in Figure 1e that the surface morphology of the powder annealed at 520 °C shows a number of pits due to degradation of phosphate [11].

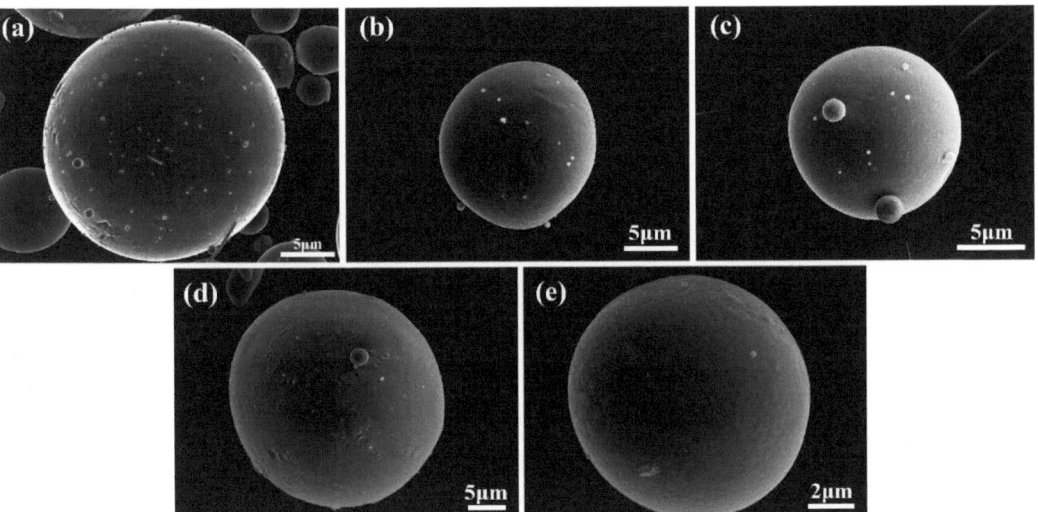

Figure 1. (a) SEM images of raw amorphous powder, (b) phosphated amorphous powder, (c–e) the phosphated powder after annealing at 440 °C, 440 °C and 520 °C, respectively.

The thermal stability of the amorphous powder was investigated. Figure 2 shows the DSC curve where only one exothermic peak at T_x = 548 °C was observed, which may correspond to the crystallization of α-Fe (Si). Details of the crystallization will be discussed in the following section.

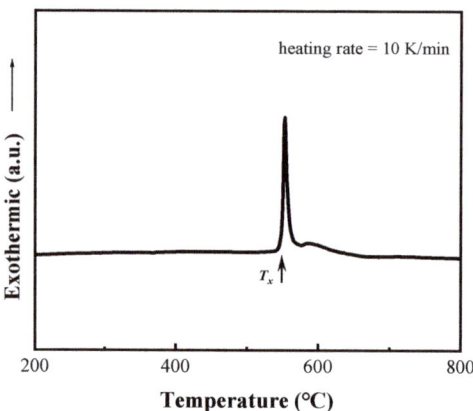

Figure 2. DSC curve of the raw FeSiBCr amorphous powder with a heating rate of 10 K/min.

The XRD patterns of phosphated powders after annealing at different temperatures for 1 h are shown in Figure 3. The patterns of the raw powder and phosphate powders are also presented for a comparison shown in Figure 3. Owing to the low concentration of phosphate, the pattern of the phosphated amorphous powders presents the same amorphous peak as the raw amorphous powders. For phosphated powders annealed at temperatures below 480 °C, no crystallization peaks can be observed. However, the crystallization peaks of the α-Fe (Si) phase appeared after annealing at 520 °C, indicating that the α-Fe (Si) phase precipitated from the amorphous matrix. Additional peaks of Fe_3B were also detected for the sample annealed at 520 °C.

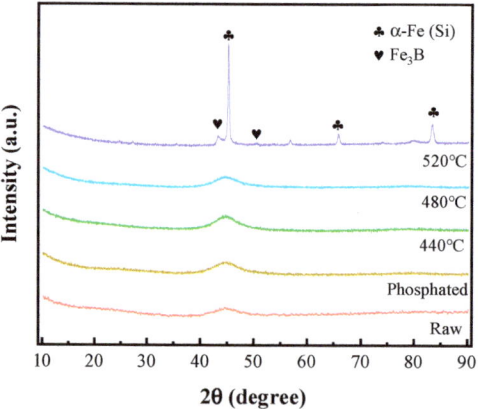

Figure 3. The XRD patterns of raw powders, phosphated powders and powders annealed at 440 °C, 480 °C and 520 °C.

Figure 4a shows the magnetization of the raw amorphous powder, phosphated powder and the powders after annealing at different temperatures. Compared with the raw powder, the saturation magnetization of phosphated powders is lower, which is attributed to the magnetic dilution of non-magnetic phosphate on the surface of the amorphous powders after phosphating [12,13]. After annealing at 440 °C, 480 °C and 520 °C, the saturation magnetizations of the composite powders are 147 emu/g, 150 emu/g and 151 emu/g.

This is due to the formation of α-Fe in the amorphous matrix and the degradation of the phosphate coating, which increases saturation magnetization [11,14]. The DC magnetic hysteresis loops of the amorphous powder cores before and after annealing at 440 °C, 480 °C and 520 °C are shown in Figure 4b. After annealing at temperatures below 480 °C, the slope of the DC hysteresis loop increases, and the area of the loop becomes smaller. As can be seen in Figure 4c, the loops in external magnetic fields of $H = -40$ to 40 Oe demonstrate that the coercivity of the unannealed amorphous powder core is 520 A/m. Meanwhile, the coercivities of the powder cores annealed at 440 °C and 480 °C are 39.3 A/m and 39.6 A/m, respectively. This is because the internal stress of the amorphous powder cores generated during cold pressing was released due to the annealing heat treatment [15]. With an increase in annealing temperature to 520 °C, the slope of the DC hysteresis loop decreases, and the loop area becomes wider than the loop of the unannealed powder cores. Combined with the results of XRD in Figure 3, after annealing at 520 °C, α-Fe and Fe_3B phases appear in the amorphous matrix. The grain boundaries generated by these crystals will hinder the movement of domain walls, which will result in a deterioration of coercivity. The corresponding saturation magnetizations and coercivities of the powder cores are shown in Figure 4d.

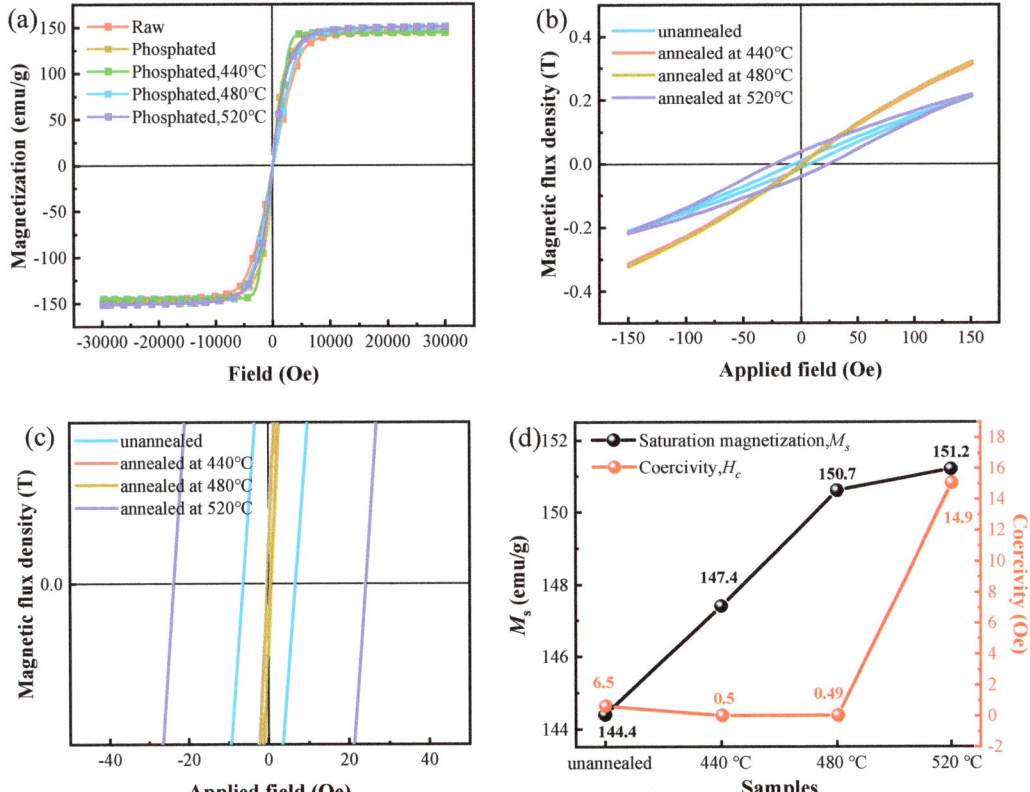

Figure 4. (**a**) M-H curves of the raw powders, coated powders and powders annealed at different temperatures; (**b**,**c**) show the DC magnetic hysteresis loop and the partial loops of the AMPCs in field $H = \pm 150$ Oe; (**d**) saturation magnetization M_s and coercivity H_c of AMPCs before and after being annealed at different temperatures.

Within the frequency range from 10 kHz to 110 MHz, the relationship between the effective permeability and frequency is illustrated in Figure 5a for the powder cores before and after annealing at different temperatures. As observed in Figure 5a, for the unannealed amorphous powder core, the effective permeability is 20.4 at low frequencies. With an increase in frequency, effective permeability presents a dispersion phenomenon [16], meaning that the effective permeability remains stable until 17 MHz. After annealing at 440 °C, the effective permeability increases to 34.3 at low frequency. With an increase in annealing temperature to 480 °C, effective permeability slightly increases to 36.4 at low frequencies. At low frequencies, the reversible magnetic domain wall movement and the reversible magnetic moment rotation contribute to effective permeability [2,7,17]. As shown in Figure 4, after annealing at 440 °C and 480 °C, the internal stress within the particles was almost all removed, which is attributed to the markedly increasing effective permeability. Although effective permeability was greatly improved, the frequency stability of effective permeability was greatly reduced. Compared with the unannealed powder core, the frequency of the effective permeability stability for powder cores annealed at 440 °C and 480 °C deteriorated to 3.2 MHz. After further increaseing the annealing temperature to 520 °C, effective permeability decreased dramatically to 20.8 and remained stable until 110 MHz. The grain boundaries within the particles annealed at 520 °C will hinder the movement of magnetic domain walls, such as internal stress, resulting in a decline in effective permeability. Figure 5b shows the frequency dependence of the quality factor Q for the powder cores after annealing at different temperatures. Usually, quality factor Q represents the efficiency of energy utilization. The Q values for the powder cores annealed at 440 °C and 480 °C have peaks of about 45.5 and 44, respectively, at 2 MHz. With an increase in annealing temperature to 520 °C, the peak value of Q decreases to 40.3, but the corresponding frequency moves to 8.5 MHz. Compared with the unannealed powder core, the powder cores annealed at 440 °C and 480 °C have larger peak values of Q, indicating that they have better energy utilization [18]. In addition, the lower the effective permeability falls, the larger the frequency corresponding to the peak value of Q, indicating that the Q-f curve shifts to the right. The overall shift of Q to the right indicates that the powder cores annealed at 440 °C and 480 °C have better frequency stability [19].

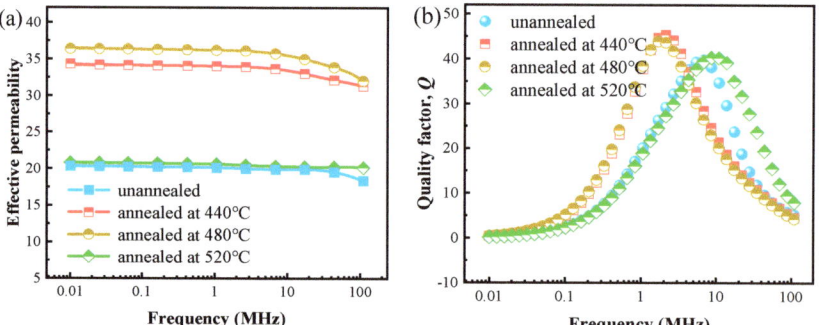

Figure 5. The magnetic properties of the AMPCs: (**a**) effective permeability μ_e versus frequency for amorphous powder cores before and after annealing at different temperatures; (**b**) quality factor Q.

Figure 6a shows core loss versus the induction of unannealed powder cores and after annealing at different temperatures at a frequency of 50 kHz. With elevating B_m, the core loss of all powder cores increased gradually, and the minimum core loss was obtained for the samples annealed at 440 °C and 480 °C. As the temperature increases to 520 °C, the core loss for the powder cores shows a noticeably higher value because it has greater coercivity due to the crystallization behavior, which hinders the movement of domain walls. Figure 6b shows the core loss in 20 kHz–800 kHz at 10 mT for all samples. With increasing annealing temperature, core losses for all powder cores decreased and then increased dramatically. The samples annealed at 440 °C and 480 °C exhibited the lowest core losses of about 55.6 mW/cm^3 and 51 mW/cm^3, respectively, at 800 kHz. In order to analyze the effect of annealing on core loss, loss separation was performed. The total core loss can be presented as follows [16,20,21]:

$$P_t = P_h + P_c = K_h \times f + K_e \times f^2 \qquad (1)$$

where K_h and K_e are the coefficients for hysteresis loss P_h and eddy current loss P_e, and f is the frequency. The total core loss has been separated and plotted in Figure 6c,d, respectively. Figure 6c shows the variation in hysteresis loss with frequency for all samples. Obviously, the hysteresis loss of all samples first decreased and then increased dramatically with increasing annealing temperature, which has the same trend as the total loss. The sample annealed at 480 °C exhibits the lowest hysteresis loss of about 29.6 mW/cm^3 at 800 kHz, and the highest hysteresis loss (approximately 213 mW/cm^3 at 800 kHz) was obtained by the sample annealed at 520 °C. Figure 6d shows the frequency dependence of eddy current loss for all powder cores. The eddy current loss for all powder cores increases with frequency in a nonlinear fashion. The unannealed sample has the highest eddy current loss of about 32 mW/cm^3 at 800 kHz, while the eddy current loss for the samples annealed at different temperatures is lower. It can be seen from Figure 6c,d that the total core loss of the unannealed sample and the sample annealed at 520 °C is dominated by the hysteresis component (213 mW/cm^3 at 800 kHz and 146 mW/cm^3 at 800 kHz, respectively) within the measurement frequency range. Owing to hysteresis loss induced by irreversible domain wall movement, internal stress and the grain boundaries generated during high-temperature annealing will hinder the motion of domain walls. Therefore, after annealing, the magnetic performance of the amorphous powder core improved.

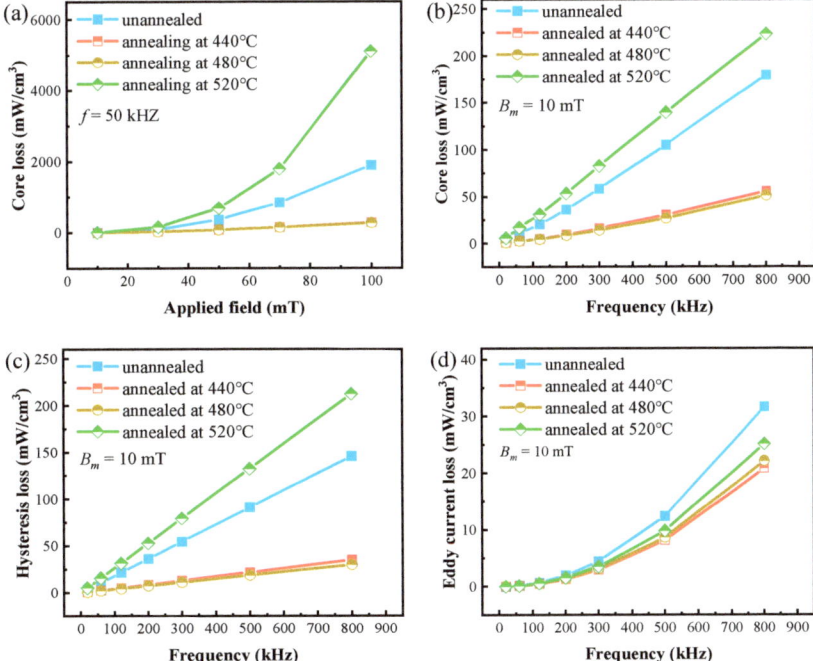

Figure 6. (**a**) Core loss versus induction of powder cores after annealing at different temperatures; (**b**) core loss versus frequency of powder cores before and after annealing at 440 °C, 480 °C and 520 °C in 20–800 kHz, at 10 mT; (**c**) hysteresis loss at 10 mT; (**d**) eddy current loss at 10 mT.

Table 1 summarizes the soft magnetic properties of the AMPCs in this paper and the typical SMCs previously reported in the literature [16,18,22–25]. It can be seen that the AMPCs in this paper maintained a relatively lower core loss of 29.6 mW/cm^3 at 800 kHz and 10 mT than the other SMCs in Table 1. Furthermore, the effective permeability of the AMPCs is about 36.4 and remains stable until 3 MHz. The improvements in permeability and core loss can expand the usage of AMPCs for high-frequency power applications.

Table 1. The core loss of the samples and other research in the literature.

Sample	μ_e	Q_{max}	P_c (mW/cm^3) 0.01 T/800 kHz	0.05 T/100 kHz	0.1 T/100 kHz	References
FeSiBCr@Phosphate	36.4	44	51.3	152	530.4	This work
FeSiBPC@Fe$_3$O$_4$@EP	49.5	160	-	187	630	[15]
FeSiBCCr@TiO$_2$	81.5	102	-	275	900	[19]
FeSiBP@(NiZn)Fe$_2$O$_4$	70	-	-	-	1000	[20]
FeSiBPNbCr@PPX	48	-	-	220	770	[21]
FeSiBP	86	-	-	200	780	[22]
FeSiCr@MnZn	48	-	45	-	-	[13]

4. Conclusions

The magnetic performance of FeSiBCr amorphous powder cores with phosphate–resin hybrid coating was significantly improved by annealing. After annealing at 440 °C and 480 °C, the internal stress within the particles was almost all released. Coercivity decreases markedly, and effective permeability increases significantly and remains stable until 3 MHz. The samples have excellent core loss at 800 kHz due to the significant reduction in hysteresis core loss, while eddy current loss remains very low after annealing below 480 °C. When the temperature increases to 520 °C, the powder cores crystallize, resulting in a deterioration of coercivity, effective permeability and core loss. The low core loss and good frequency stability of the amorphous powder cores provide broad prospects for their application in electronic components at high frequencies.

Author Contributions: Conceptualization, H.Y., J.L. (Jiaming Li), J.L. (Jingzhou Li), X.C., G.H., J.Y., R.C.; methodology and data curation, J.L. (Jiaming Li); writing—original draft preparation, J.L. (Jiaming Li); supervision, H.Y. All authors have read and agreed to the published version of the manuscript.

Funding: This work was supported by the Dongguan Innovative Research Team Program (Grant No. 2020607231010) and Guangdong Provincial Natural Science Foundation of China (No. 2021A1515010642).

Conflicts of Interest: The authors declare no conflict of interest.

References

1. Perigo, E.A.; Weidenfeller, B.; Kollár, P.; Füzer, J. Past, present, and future of soft magnetic composites. *Appl. Phys. Rev.* **2018**, *5*, 031301. [CrossRef]
2. Silveyra, J.M.; Ferrara, E.; Huber, D.L.; Monson, T.C. Soft magnetic materials for a sustainable and electrified world. *Science* **2018**, *362*, eaao0195. [CrossRef] [PubMed]
3. Gutfleisch, O.; Willard, M.A.; Brück, E.; Chen, C.H.; Sankar, S.G.; Liu, J.P. Magnetic materials and devices for the 21st century: Stronger, lighter, and more energy efficient. *Adv. Mater.* **2011**, *23*, 821–842. [CrossRef] [PubMed]
4. Chen, Z.; Liu, X.; Kan, X.; Wang, Z.; Zhu, R.; Yang, W. Phosphate coatings evolution study and effects of ultrasonic on soft magnetic properties of FeSiAl by aqueous phosphoric acid solution passivation. *J. Alloys Compd.* **2019**, *783*, 434–440. [CrossRef]
5. Wang, C.; Liu, J.H.; Peng, X.L.; Li, J.; Yang, Y.T.; Han, Y.B.; Xu, J.C.; Gong, J.; Hong, B.; Ge, H.L.; et al. FeSiCrB amorphous soft magnetic composites filled with Co$_2$Z hexaferrites for enhanced effective permeability. *Adv. Powder Technol.* **2022**, *33*, 103378. [CrossRef]
6. Pittini-Yamada, Y.; Périgo, E.A.; De Hazan, Y.; Nakahara, S. Permeability of hybrid soft magnetic composites. *Acta Mater.* **2011**, *59*, 4291–4302. [CrossRef]
7. Shokrollahi, H.; Janghorban, K. Soft magnetic composite materials (SMCs). *J. Mater. Process. Technol.* **2007**, *189*, 1–12. [CrossRef]

8. Hossain, M.K.; Ferdous, J.; Haque, M.M.; Hakim, A.A. Development of nanostructure formation of $Fe_{73.5}Cu_1Nb_3Si_{13.5}B_9$ alloy from amorphous state on heat treatment. *World J. Nano Sci. Eng.* **2015**, *5*, 107–114. [CrossRef]
9. Hossain, M.K.; Ferdous, J.; Haque, M.M.; Hakim, A.A. Study and characterization of soft magnetic properties of $Fe_{73.5}Cu_1Nb_3Si_{13.5}B_9$ magnetic ribbon prepared by rapid quenching method. *Mater. Sci. Appl.* **2015**, *6*, 1089–1099.
10. Li, W.; Xiao, S.; Li, W.; Ying, Y.; Yu, J.; Zheng, J.; Qiao, L.; Li, J.; Naoki, W.; Wu, J.; et al. Hybrid amorphous soft magnetic composites with ultrafine FeSiBCr and submicron FeBP particles for MHz frequency power applications. *J. Magn. Magn. Mater.* **2022**, *555*, 169365. [CrossRef]
11. Ouda, K.; Danninger, H.; Gierl-Mayer, C.; Hellein, R.; Müller, A. Ferrothermal reduction of iron(III)phosphate insulating layers in soft magnetic composites. *Powder Metall.* **2021**, *64*, 351–359. [CrossRef]
12. Liu, D.; Wu, C.; Yan, M.; Wang, J. Correlating the microstructure, growth mechanism and magnetic properties of FeSiAl soft magnetic composites fabricated via HNO_3 oxidation. *Acta Mater.* **2018**, *146*, 294–303. [CrossRef]
13. Xia, C.; Peng, Y.; Yi, Y.; Deng, H.; Zhu, Y.; Hu, G. The magnetic properties and microstructure of phosphated amorphous FeSiCr/silane soft magnetic composite. *J. Magn. Magn. Mater.* **2019**, *474*, 424–433. [CrossRef]
14. Li, F.C.; Liu, T.; Zhang, J.Y.; Shuang, S.; Wang, Q.; Wang, A.D.; Wang, J.G.; Yang, Y. Amorphous–nanocrystalline alloys: Fabrication, properties, and applications. *Mater. Today Adv.* **2019**, *4*, 100027. [CrossRef]
15. Zhou, B.; Chi, Q.; Dong, Y.; Liu, L.; Zhang, Y.; Chang, L.; Pan, Y.; He, J.; Li, J.; Wang, X. Effects of annealing on the magnetic properties of Fe-based amorphous powder cores with inorganic-organic hybrid insulating layer. *J. Magn. Magn. Mater.* **2020**, *494*, 165827. [CrossRef]
16. Nie, W.; Yu, T.; Wang, Z.; Wei, X. High-performance core-shell-type FeSiCr@MnZn soft magnetic composites for high-frequency applications. *J. Alloys Compd.* **2021**, *864*, 158215. [CrossRef]
17. Dobák, S.; Füzer, J.; Kollár, P.; Fáberová, M.; Bureš, R. Interplay of domain walls and magnetization rotation on dynamic magnetization process in iron/polymer–matrix soft magnetic composites. *J. Magn. Magn. Mater.* **2017**, *426*, 320–327. [CrossRef]
18. Chi, Q.; Chang, L.; Dong, Y.; Zhang, Y.; Zhou, B.; Zhang, C.; Pan, Y.; Li, Q.; Li, J.; He, A.; et al. Enhanced high frequency properties of FeSiBPC amorphous soft magnetic powder cores with novel insulating layer. *Adv. Powder Technol.* **2021**, *32*, 1602–1610. [CrossRef]
19. Wang, J.; Wu, Z.; Li, G. Intergranular insulated Fe/SiO_2 soft magnetic composite for decreased core loss. *Adv. Powder Technol.* **2016**, *27*, 1189–1194. [CrossRef]
20. Li, W.; Cai, H.; Kang, Y.; Ying, Y.; Yu, J.; Zheng, J.; Qiao, L.; Jiang, Y.; Che, S. High permeability and low loss bioinspired soft magnetic composites with nacre-like structure for high frequency applications. *Acta Mater.* **2019**, *167*, 267–274. [CrossRef]
21. Guo, Z.; Wang, J.; Chen, W.; Chen, D.; Sun, H.; Xue, Z.; Wang, C. Crystal-like microstructural Finemet/FeSi compound powder core with excellent soft magnetic properties and its loss separation analysis. *Mater. Des.* **2020**, *192*, 108769. [CrossRef]
22. Zhou, B.; Dong, Y.; Liu, L.; Chi, Q.; Zhang, Y.; Chang, L.; Bi, F.; Wang, X. The core-shell structured Fe-based amorphous magnetic powder cores with excellent magnetic properties. *Adv. Powder Technol.* **2019**, *30*, 1504–1512. [CrossRef]
23. Xiaolong, L.; Yaqiang, D.; Min, L.; Chuntao, C.; Xin-Min, W. New Fe-based amorphous soft magnetic composites with significant enhancement of magnetic properties by compositing with nano-$(NiZn)Fe_2O_4$. *J. Alloys Compd.* **2017**, *696*, 1323–1328. [CrossRef]
24. Zhang, Y.; Dong, Y.; Zhou, B.; Chi, Q.; Chang, L.; Gong, M.; Huang, J.; Pan, Y.; He, A.; Li, J.; et al. Poly-para-xylylene enhanced Fe-based amorphous powder cores with improved soft magnetic properties via chemical vapor deposition. *Mater. Des.* **2020**, *191*, 108650. [CrossRef]
25. Chang, C.; Dong, Y.; Liu, M.; Guo, H.; Xiao, Q.; Zhang, Y. Low core loss combined with high permeability for Fe-based amorphous powder cores produced by gas atomization powders. *J. Alloys Compd.* **2018**, *766*, 959–963. [CrossRef]

Article

Properties Optimization of Soft Magnetic Composites Based on the Amorphous Powders with Double Layer Inorganic Coating by Phosphating and Sodium Silicate Treatment

Pan Luo [1], Hongya Yu [1,*], Ce Wang [1], Han Yuan [1], Zhongwu Liu [1,*], Yu Wang [2], Lu Yang [2] and Wenjie Wu [2]

1. School of Materials Science and Engineering, South China University of Technology, Guangzhou 510640, China
2. Southwest Institute of Applied Magnetism, Mianyang 621000, China
* Correspondence: yuhongya@scut.edu.cn (H.Y.); zwliu@scut.edu.cn (Z.L.)

Citation: Luo, P.; Yu, H.; Wang, C.; Yuan, H.; Liu, Z.; Wang, Y.; Yang, L.; Wu, W. Properties Optimization of Soft Magnetic Composites Based on the Amorphous Powders with Double Layer Inorganic Coating by Phosphating and Sodium Silicate Treatment. Metals 2023, 13, 560. https://doi.org/10.3390/met13030560

Academic Editor: Joan-Josep Suñol

Received: 14 February 2023
Revised: 5 March 2023
Accepted: 8 March 2023
Published: 10 March 2023

Copyright: © 2023 by the authors. Licensee MDPI, Basel, Switzerland. This article is an open access article distributed under the terms and conditions of the Creative Commons Attribution (CC BY) license (https://creativecommons.org/licenses/by/4.0/).

Abstract: Core-shell structured amorphous FeSiBCr@phosphate/silica powders were prepared by phosphating and sodium silicate treatment. The soft magnetic composites (SMCs) were fabricated based on these powders. The effects of phosphoric acid (H_3PO_4) concentration and annealing temperature on their properties were investigated. During the phosphating process, the powder coated with a low concentration of H_3PO_4-ethanol solution leads to uneven phosphate coating, while the peeling of phosphate coating occurs for the high H_3PO_4 concentration. Using 0.5 wt.% phosphoric solution, a uniform and dense insulation layer can be formed on the surface of the powder, resulting in increased resistivity and the reduced eddy current loss of the amorphous soft magnetic composites (ASMCs). This insulation layer can increase the roughness of the powder surface, which is beneficial to the subsequent coating of sodium silicate. By optimizing sodium silicate treatment, a complete and uniform SiO_2 layer can be formed on the phosphated powders well, leading to double layer core-shell structure and excellent soft magnetic properties. The magnetic properties of amorphous SMCs can be further improved by post annealing due to the effectively released residual stress. The enhanced permeability and greatly reduced core loss can be achieved by annealing at 773 K, but the deterioration of magnetic properties occurs as the annealing temperature over 798 K, mainly due to the increase of α-Fe(Si) and Fe_3B phases, which hinder the domain wall displacement and magnetic moment rotation. The excellent soft magnetic properties with permeability μ_e = 35 and core loss P_s = 368 kW/m^3 at 50 mT/200 kHz have been obtained when the SMCs prepared with the powders coated by 0.5 wt.% H_3PO_4 and 2 wt.% sodium silicate were annealed at 773 K.

Keywords: soft magnetic composites; amorphous powder; phosphating; sodium silicate; annealing; magnetic properties

1. Introduction

Soft magnetic composites (SMCs), consisting of soft magnetic powders and insulation binder, have found increasing applications in various electronic and electrical components due to their stable permeability and low core loss at high frequency [1–4]. With the development of magnetic devices towards miniaturization, high frequency, and high efficiency, SMCs with high magnetic saturation, stable permeability, and low core loss are urgently required [5–7]. Currently, there are various types of magnetic powders for SMCs, including pure iron (Fe), Fe-Si, Sendust (Fe-Si-Al), Flux (Fe-Ni), MPP (Fe-Ni-Mo), Fe-based amorphous, and nanocrystalline alloys. The Fe-based amorphous alloys such as FeSiBCr prepared by gas atomization exhibit high potential to work as the ideal core materials due to their good sphericity, low coercivity, and high saturation magnetization.

The preparation of SMCs involves coating an insulation layer on the metallic powders followed by powder forming and curing. Both inorganic coatings and organic coatings are employed. Inorganic coatings including oxides (SiO_2 [8–10], MgO [11], TiO_2 [12],

Al_2O_3 [13], ZrO_2 [14]), ferrites (MnZn ferrite [15], NiZn ferrite [16]) and inorganic salts exhibit high temperature resistance and high electric resistivity. They are generally prepared by in-situ chemical synthesis or coating a mixture of inorganic nanoparticles and organic resin on the powder. However, it was found that, for the oxide coating, a uniform and dense layer tightly bound to the magnetic powder is difficult to prepare. Soft magnetic ferrites can be used for coating, which exhibits high resistivity and high magnetization, but the brittle ferrite layer is easy to break during the pressing process. The inorganic salt layer can be prepared by direct reaction of the acid with the magnetic powder [15]. These coatings have the advantages of high density, high uniformity, high resistivity, and strong bonding with the magnetic powder. The common acids include HNO_3 [17,18] and H_3PO_4 [19,20], among which H_3PO_4 is frequently used for industrial production. However, for the amorphous Fe-based alloy powders, it is difficult to prepare an insulating layer on their surfaces by H_3PO_4 treatment [21] since amorphous alloys generally show excellent corrosion resistance [22].

It has been reported [8] that silica (SiO_2) coating can work as effective high-resistive thin layer for the magnetic powders. Sol-gel method is generally employed for preparing SiO_2 coating on Fe-based powder, but this process is complicated since four kinds of chemicals including tetraethoxysilane (TEOS) are needed [23]. Recently, FeSiCr powder was coated with a SiO_2 layer by reacting carbon dioxide with sodium silicate, which indicates that sodium silicate (Na_2SiO_3) can be used as a silicon source [24]. However, the insulating layer prepared by sodium silicate is not dense (in particular for the atomized amorphous powder with the smooth surface).

In this work, in order to obtain a uniform and dense insulation layer on the amorphous Fe-based alloy powder, an inorganic double layer coating is proposed, which was prepared by phosphating followed by SiO_2 coating. The phosphating layer by H_3PO_4 treatment is employed to increase the surface roughness of the powder, which is beneficial for depositing a uniform and dense insulation SiO_2 layer. By this approach, we prepared core-shell structured amorphous FeSiBCr@phosphate/SiO_2 powders, which then were made into magnetic cores. The effects of H_3PO_4 concentration and annealing temperature on the soft magnetic properties of SMCs were investigated. This method is suitable for the industrial production of amorphous SMCs with stable permeability and low core loss.

2. Experimental Procedures

Commercial gas atomized $Fe_{87}Si_{5.5}B_4Cr_{3.5}$ amorphous powders were supplied by Tiz-Advanced Alloy Technology Co., Ltd (Quanzhou, China). The average particle size D50 of the raw powders is approximately 20 μm and the *apparent density of powders* is 3.5–4.0 g/cm^3. The saturation magnetization (M_s) of the raw powders is 148 Am^2/kg. Sodium silicate (Na_2SiO_3) with modulus of 1 is purchased from Guangzhou Chemical Reagent Factory (Guangzhou, China). To prepare SMCs, the FeSiBCr powders (30 g) were first dispersed in H_3PO_4-ethanol solution (30 g) with various H_3PO_4 concentrations (0 wt.%, 0.25 wt.%, 0. 5 wt.%, 0.75 wt.%, 1.0 wt.%) under constant stirring at 55 °C until the ethanol solvent was completely evaporated for surface passivation. After washing and drying, the phosphated powders were obtained. The phosphatized powders were then dispersed in 2 wt.% sodium silicate aqueous solution (15 g) and stirred at 328 K for 20 min. The obtained magnetic powders were filtered and washed with deionized water for three times. The treated powders were mixed with 2.5 wt.% silicone resin and 0.5 wt.% epoxy resin-acetone solution and stirred at room temperature until the acetone solvent evaporated completely for binding. The bonded powders were compacted into toroidal-shaped amorphous SMCs with the dimensions of $\Phi 20 \times \Phi 12 \times 4$ mm under 1200 MPa for 5.5 s. The SMCs were cured at 453 K for 1 h. For annealing treatment, the selected amorphous SMCs were heated at different temperatures (748, 773, 798 K) for 1 h under Ar atmosphere.

The surface morphologies of the powders were observed by scanning electron microscopy (SEM, FEI Quanta 200, Hillsboro, OR, USA). Fourier transform infrared spectrometer (FTIR, VERTEX 70, Bruker, Billerica, MA, USA) was used to investigate the phase constitution of the insulation layer. The soft magnetic properties were obtained by LCR meter (IM3536, Hioki, Nagano, Japan) and soft magnetic AC measurement device (MATS-3010SA, Linkjoin, Loudi, China). The electrical resistivity was measured by a high resistance weak current tester (ST2643, Jingge, SuZhou, China). The phase structures of powders were confirmed by X-ray diffraction (XRD, Philips X' Pert, PANalytical, Almelo, The Netherlands) using Cu Kα radiation at room temperature in the range of 20–80°. The magnetic hysteresis loops were measured at room temperature using a vibrating sample magnetometer (VSM, East Changing, Beijing, China). The radial crush strength was measured by core rupture test machine (FL-8621, Feiling, Dongguan, China).

3. Results and Discussion

3.1. The Formation of Double Layer Structure on the Powder

Figure 1 shows the SEM images of the amorphous FeSiBCr powders before and after different treatments. In Figure 1a, the original powder shows smooth surface, which is unfavorable for the coating of sodium silicate through physical contact. The surface of the powder after phosphatized in a 0.5 wt.% H_3PO_4-ethanol solution becomes rough, as shown in Figure 1b. The rough phosphate structure can improve the adhesion of sodium silicate. After treated only by sodium silicate, the powder surface exhibits loose and porous structure (Figure 1c). A dense insulating layer can be formed on the surface of the powder after two-step treatment by both H_3PO_4 and sodium silicate, as shown in Figure 1d. As we know, the uniform and dense insulation layer on the powders is beneficial to the soft magnetic properties of SMCs.

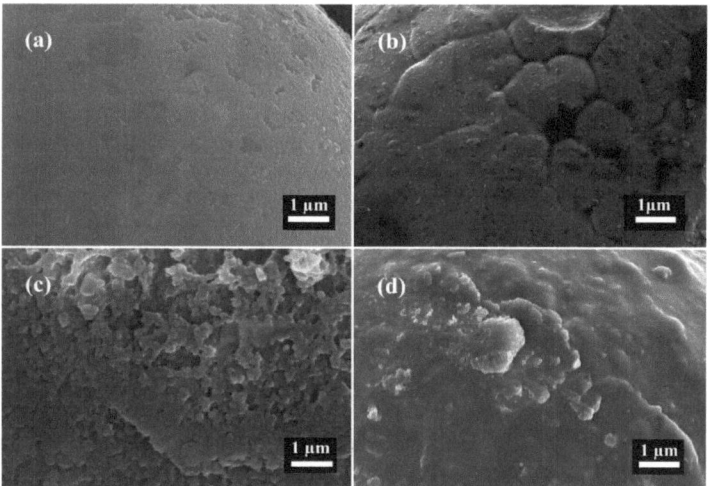

Figure 1. SEM images of (**a**) the original powder and the powders treated by (**b**) 0.5 wt.% H_3PO_4-ethanol solution, (**c**) 2 wt.% sodium silicate-aqueous solution, and (**d**) 0.5 wt.% H_3PO_4-ethanol solution followed by 2 wt.% sodium silicate-aqueous solution.

FTIR analysis results for the amorphous FeSiBCr powder and the powders after different treatments are shown in Figure 2. FTIR Curve (a) for the original powders shows no distinct band. Bands of 2915 cm^{-1} (Curve (b) and (c)) are attributed to the-CH_3 group from residual C_2H_5OH. The characteristic peaks of phosphate are observed at the band of 1065 cm^{-1} and 540 cm^{-1} for the powders treated by H_3PO_4-ethanol solution (Curve (b)). The bands at 1640 cm^{-1} and 1065 cm^{-1} are assigned to the bending mode of P-OH bonds and the symmetric stretching vibrations of P-O bonds [25], respectively. The results indicate that the phosphate has been formed on the surface of the magnetic powder after H_3PO_4 treatment. On Curve (c) for the powders coated by sodium silicate-aqueous solution, the broad absorption band at 1038 cm^{-1} originates from the symmetric vibration absorption of Si-O-Si [26], which confirms that the obtained coating layer on the powders is SiO$_2$. On Curve (d) for the powders treated with H$_3$PO$_4$-ethanol followed by sodium silicate solution, the characteristic bonds of both phosphate and SiO$_2$ can be detected, indicating that both phosphate and SiO$_2$ layers are formed on the powders. Compared with phosphate and SiO$_2$ coating alone, the hybrid phosphate-SiO$_2$ layer exhibits significantly decreased intensity of the absorption peak corresponding to -OH at 3450 cm^{-1}, indicating that the -OH of phosphate binds to the -OH of sodium silicate. The results thus suggest that the P-O-Si bond may be formed on the powder, which can enhance the binding of SiO$_2$ and magnetic powder.

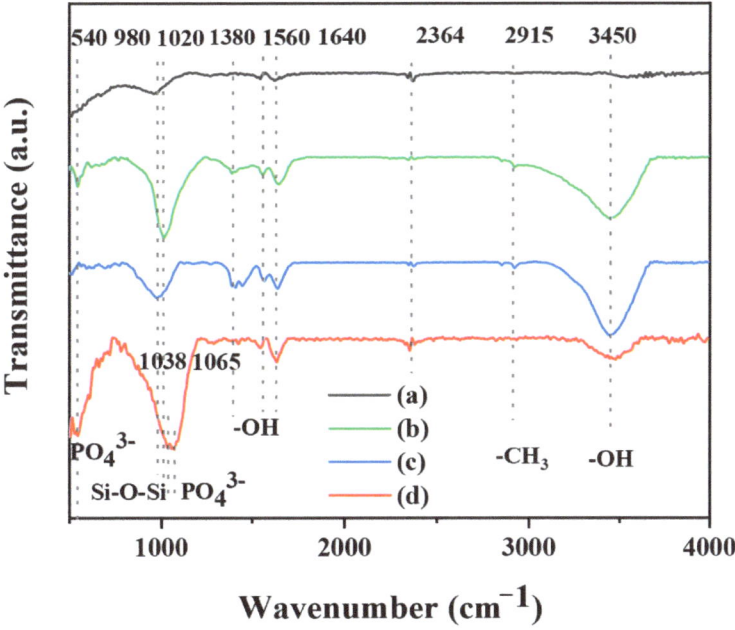

Figure 2. FTIR curves of (**a**) the original FeSiBCr powders, and the powders treated by (**b**) H$_3$PO$_4$-ethanol solution, (**c**) sodium silicate-aqueous solution, and (**d**) H$_3$PO$_4$-ethanol solution followed by sodium silicate-aqueous solution.

The evolution of the core-shell structure on the magnetic powder can be illustrated in Figure 3. In the first step 1, H$_3$PO$_4$ reacts with the FeSiBCr powder according to Equation (1), and the first layer of phosphate can be formed on the powder [27,28]. In the second step, Si(OH)$_4$ dehydrates and condenses with the -OH on the phosphated powder surface to form a networked SiO$_2$, and the second shell on the powder is formed. Hydrolysis reaction and condensation reaction can be expressed by Equations (2) and (3) [29]. As a result, the core-shell structured FeSiBCr@phosphate/SiO$_2$ powders can be prepared.

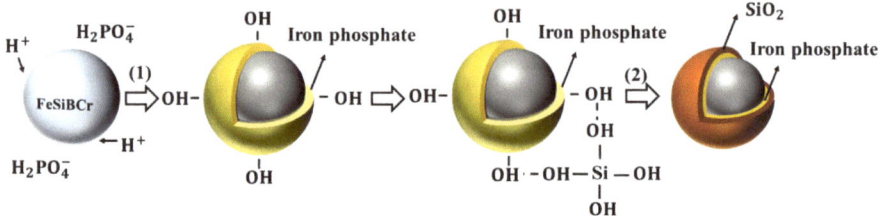

Figure 3. Evolution of the core-shell structure of magnetic powder. (**1**) Phosphate coating process; (**2**) SiO$_2$ coating process.

$$Fe^{2+} + HPO_4^{2-} \rightarrow FeHPO_4$$
$$3Fe^{2+} + 2PO_4^{3-} \rightarrow Fe_3(PO_4)_2 \quad (1)$$

$$4Fe^{2+} + O_2 + 4PO_4^{3-} + 4H^+ \rightarrow 4FePO_4 + 2H_2O$$
$$Na_2O \cdot nSiO_2 + (2n+1)H_2O \rightarrow 2NaOH + nSi(OH)_4 \quad (2)$$

$$nSi(OH)_4 \rightarrow [Si(OH)_4]n \xrightarrow{-2H_2O} \begin{array}{c} | \quad\quad | \\ -Si-O-Si- \\ | \quad\quad | \\ O \quad\quad O \\ | \quad\quad | \\ -Si-O-Si- \\ | \quad\quad | \end{array} \quad (3)$$

3.2. Effects of Phosphated by Various H$_3$PO$_4$ Concentrations

Figure 4a–d show the SEM images of the powders phosphated by the H$_3$PO$_4$ with different concentrations (0.25 wt.%, 0.5 wt.%, 0.75 wt.%, 1.0 wt.%). For the powder phosphated with 0.25 wt.% H$_3$PO$_4$, P and O elements are evenly distributed on the powder surface, as shown in Figure 4a with EDS result. It is believed that a phosphating layer is formed. Comparing the morphologies in Figure 4a–d, the higher concentration of H$_3$PO$_4$ leads to the rougher surface of the powder. When the H$_3$PO$_4$ concentration increases to 1 wt.%, a large amount of sheet structures appears. EDS testing shows that these sheet structures contain abundant P, O, and Fe elements, presumably phosphate, as shown in Figure 4d. The appearance of the sheet structured phosphate may destroy the integrity of the layer and reduce the density of the SMCs. Thus, it is necessary to control the nucleation and precipitation of the flaky phosphate to make the phosphate coating strong and dense.

Table 1 shows the effective permeability μ_e, quality factor Q and core loss P_s of the magnetic cores prepared from the powders treated with 0.5 wt.% H$_3$PO$_4$-ethanol solution followed by different concentrations of sodium silicate solution. All cores were annealed at 773 K for 1 h. It was found that an appropriate amount of SiO$_2$ can help to form a uniform and complete insulation layer on the powders, which will not completely break during the pressing process. However, more SiO$_2$ with high hardness may result in decreased density and deteriorated magnetic properties [30]. The core based on powders experienced 2 wt.% sodium silicate solution offer the optimal performance with high permeability, high Q, and low magnetic loss. Therefore, in the following experiments, for studying the effects of H$_3$PO$_4$ concentration on the properties of SMCs, 2 wt.% sodium silicate-aqueous solution were selected for coating SiO$_2$.

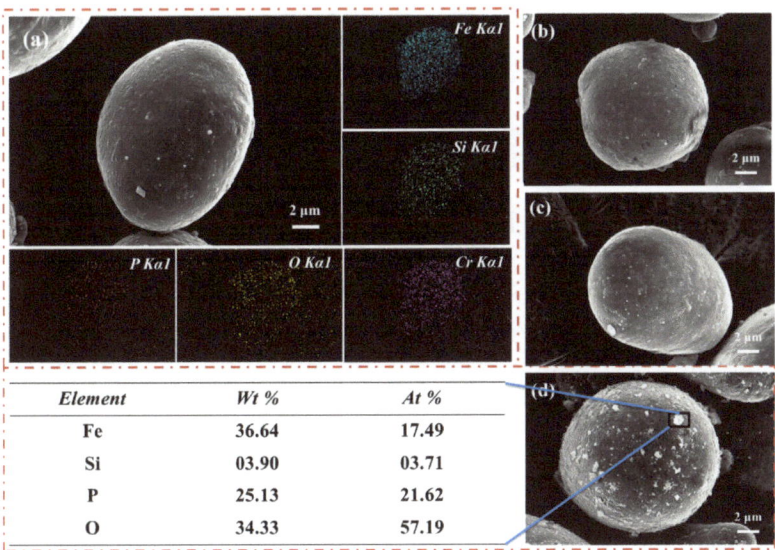

Figure 4. SEM images of the powders phosphated by different H_3PO_4 concentrations (**a**) 0.25 wt.%, (**b**) 0.5 wt.%, (**c**) 0.75 wt.%, and (**d**) 1.0 wt.%. The EDS result of powder phosphated by H_3PO_4 concentration of 0.25 wt.% is also shown in the figure.

Table 1. Effective permeability μ_e, Q and core loss P_s of the SMCs with 0.5 wt.% H_3PO_4-ethanol solution and different concentration of sodium silicate-aqueous solution.

Sample (Powder Treatment)	Effective Permeability (μ_e)	Q	P_s (kW/m^3)	
			50 mT/100 kHz	50 mT/200 kHz
Phosphatized 0.5 wt.%	37.8	42	301.0	703.6
Phosphatized 0.5 wt.% + Sodium silicate 1 wt.%	36.4	35	243.6	576.3
Phosphatized 0.5 wt.% + Sodium silicate 2 wt.%	33.4	50	136.4	368.0
Phosphatized 0.5 wt.% + Sodium silicate 3 wt.%	32.0	45	187.1	490.3
Phosphatized 0.5 wt.% + Sodium silicate 4 wt.%	29.0	35	194.3	564.6

Figure 5 shows the magnetic properties of SMCs fabricated by the amorphous FeSiBCr powders treated with various H_3PO_4-ethanol solutions followed by 2 wt.% sodium silicate-aqueous solution and annealed at 773 K. For all samples, the permeability decreases as the frequency increases, mainly due to the weakening effect of eddy currents generated by induced electromotive force on the applied magnetic field at high frequencies (Figure 5a). It decreases from ~37.8 to ~29.0 with increasing H_3PO_4 concentration from 0 to 1 wt.%. Increasing H_3PO_4 concentration leads to more non-magnetic material, resulting in lower permeability. When the H_3PO_4 concentration is 0.5 wt.%, the SMCs has the beat frequency stability, as shown in Figure 2. The insulating layer prepared by 0.5 wt.% H_3PO_4 concentration followed by 2 wt.%Na_2SiO_3 has the beat coating effect and can effectively isolate eddy currents between powders.

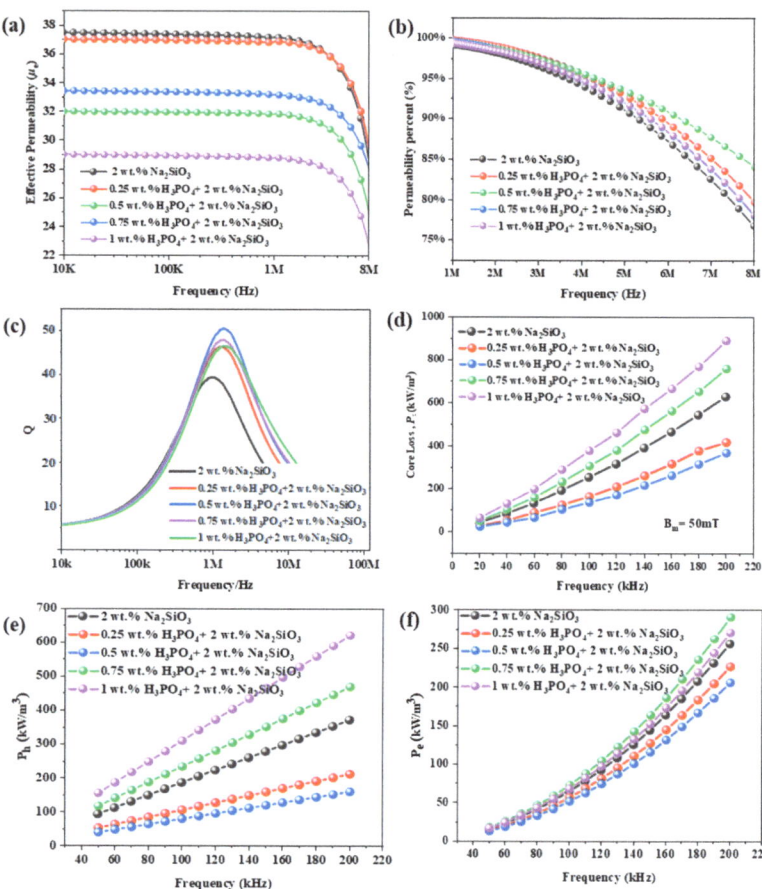

Figure 5. Frequency dependences of the (**a**) μ_e, (**b**) permeability percent, (**c**) Q, (**d**) P_s, (**e**) P_h, (**f**) P_e for SMCs after coating with different H_3PO_4 concentrations.

Q value is the quality factor, which is an important indicator to measure the high-frequency performance of soft magnetic composites. The Q value can be expressed by the ratio of the real permeability μ of the soft magnetic composite to the imaginary permeability μ', as well as the ratio of the inductance and the equivalent loss resistance, expressed as Equation (4):

$$Q = \frac{\mu'}{\mu''} = \left(\frac{2\pi f L_s}{R_s}\right) \qquad (4)$$

where L_s is the inductance and R_s is the equivalent loss resistance. In general, the higher the Q value on behalf of powder cores, the lower the rate of energy loss. The Q value of the sample first increases then decreases with increasing frequency (Figure 5b). The SMCs treated with 0.5 wt.% H_3PO_4 concentration exhibits a maximum peak value of 50, which shows a significant performance improvement compared to those without H_3PO_4 treatment.

The frequency dependence of core loss (P_s) for the SMCs prepared with different H_3PO_4 concentrations in the range of 20 kHz to 200 kHz at B_m = 50 mT are shown in Figure 5c. The P_s of all samples increase rapidly with increasing frequency. The lowest P_s of 368 kW/m^3 at 200 kHz is obtained for the SMC with 0.5 wt.% H_3PO_4 concentration, decreased by 41% compared to that without phosphating. P_s decreases from 628 to

368 kW/m³ until the H_3PO_4 concentration increases from 0 to 0.5 wt.%. As the concentration of H_3PO_4 exceeds 0.5 wt.%, the P_s shows a rapid increase. As we know, P_s is mainly determined by the hysteresis loss (P_h), the eddy loss (P_e) and residual loss (P_r), which can be expressed by Equation (5) [31]:

$$P_{s\ (total)} = P_h + P_e + P_r \approx f \oint HdB + \frac{CB^2d^2f^2}{\rho} + K_r Bxf^{1.5} \tag{5}$$

$$P_e = P_e^{Intra} + P_e^{Inter} = \frac{(\pi d_{powder} B_m)^2}{20\rho_{powder} R_{powder}} f^2 + \frac{(\pi d_{eff} B_m)^2}{\beta \rho_s R_s} f^2 \tag{6}$$

where H is the magnetic field, B is the magnetic flux density, f is the frequency, C is the constant, d is the effective diameter of powders and ρ is the resistivity. K_r is residual loss coefficient related to the material, and x is the coefficient related to the magnitude and frequency of the applied magnetic field. P_r is a combination of relaxation and resonant losses. These losses are only important at very low induction levels and very high frequencies and can be ignored in power applications [1,32,33]. Eddy current losses include intra-particle and inter-particle eddy current losses, as shown in Equation (6), d_{powder} is the effective size of the powder, ρ_{powder} is the density of the powder, and R_{powder} is the resistivity of the powder. d_{eff} is the effective size of the eddy current, ρ_s is the density of the SMCs, and R_s is the resistivity of the SMCs. P_e^{Intra} is mainly related to the resistivity and particle size of the powder. P_e^{Inter} is mainly related to the degree of insulation between powder particles, that is, the resistivity of SMCs. In this paper, by using different insulation coating processes, the Rs of the SMCs can be effectively improved, thereby reducing the P_e^{Inter}.

P_s can be separated into P_h and P_e based on above equation and the results are presented in Figures 5d and 5e, respectively. When the H_3PO_4 concentration exceeds 0.5 wt.%, P_h and P_e increase with the increase of H_3PO_4 concentration. This can be understood since the excess iron phosphate precipitates between the soft magnetic composite particles [34]. Sodium silicate can react with dehydration and condensation on the detached phosphate layer, which worsens the coating effect of the magnetic powder, leading to the increase of P_h and P_e. The reduced P_e is mainly due to the electrical resistivity reduction resulting from the destruction of the layer, as shown in Figure 6.

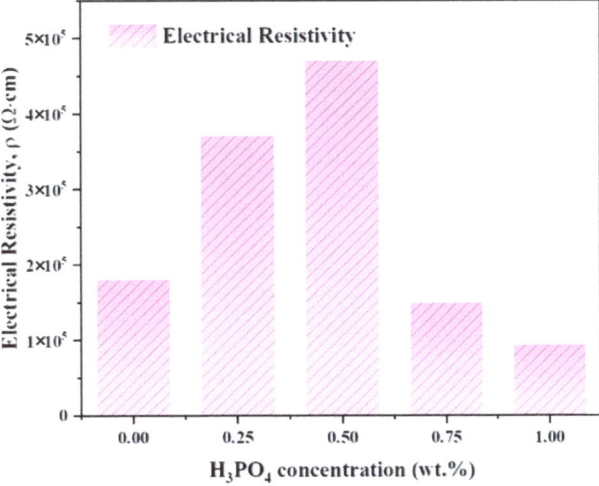

Figure 6. Electrical resistivity for SMCs after coating with different H_3PO_4 concentrations.

The magnetic composite consists of core and surface layer. For a spherical magnetic particle coated with a layer, the saturation magnetization (M_S) can be expressed as Equation (7),

$$M_S = \frac{M_{SP} \cdot V_P + M_{SL} \cdot V_L}{V_P + V_L} = M_{SP} - \frac{6t}{d}(M_{SP} - M_{SL}) \tag{7}$$

where d and t represent the diameter of magnetic particle and the thickness of surface layer, respectively. M_{SP} and M_{SL} are the saturation magnetizations of the particle and layer, respectively. V corresponds to the volume [35]. The hysteresis loops of FeSiBCr powders treated by different processes are recorded in Figure 7. The powders show a typical characteristic of soft magnetic material, including a high saturated magnetization and a low coercivity. The maximum M_s of raw FeSiBCr powders is 148 Am^2/kg. After treated by H$_3$PO$_4$ (0.5 wt.%) and sodium silicate (2 wt.%), the M_s decreases to 145 Am^2/kg and to 137 Am^2/kg, respectively. However, when coated with phosphate followed by SiO$_2$, M_s is 143 Am^2/kg, and this value is somewhere between the phosphate and SiO$_2$ coating alone. Compared with the layer of SiO$_2$, the layer of hybrid phosphate and SiO$_2$ has a lower M_{SL} due to the increase of non-magnetic substances. The increased M_S value for the powders coated with phosphate and SiO$_2$ indicated that the thickness t of the phosphate-SiO$_2$ layer is less than that of SiO$_2$ layer according to Equation (5). The results show that the double layer coating has a higher density, which gives a smaller thickness. As a result, the SMC prepared with core-shell structured FeSiBCr@phosphate/SiO$_2$ powders exhibits the best soft magnetic properties.

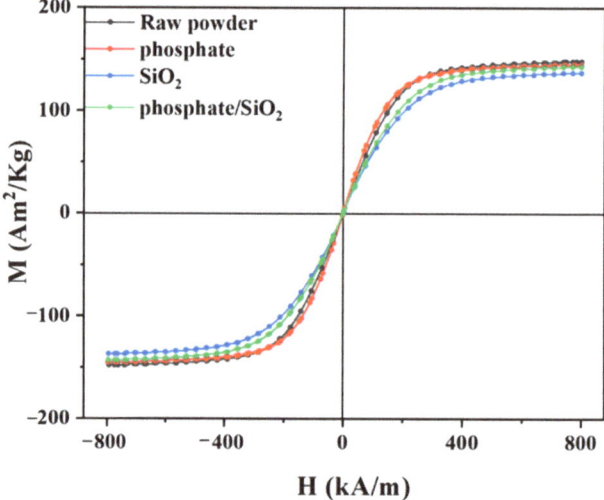

Figure 7. Magnetic hysteresis curves of FeSiBCr powders for different treatment processes.

3.3. Effects of Annealing Treatment

Figure 8a shows the variation of the effective permeability (μ_e) with frequency for SMCs treated by H$_3$PO$_4$ (0.5 wt.%) and sodium silicate (2 wt.%) followed by annealing at different temperatures. With the increasing annealing temperature from 748 to 773 K, the μ_e increases from 28.1 to 33.4. In general, the increased annealing temperature can decrease the internal stress of magnetic powder core, thereby reducing the difficulty of magnetic domain reversal and increasing the permeability [36]. The internal stress for SMCs after annealing at 748 K is released incompletely. With further elevating the annealing temperature to 773 K, the internal stress within the particles was almost removed, leading to the increase of μ_e. When the annealing temperature increases to 798 K, the μ_e of the SMCs decreases sharply to 28.0. For the SMCs annealed at 773 K, the μ_e reaches the highest value of 33.4, increased by 43%, compared with the μ_e of 18.9 for that without annealing treatment. Similarly, due to the stress release, the quality factor (Q) shows a peak value of 50 at a frequency of 3 MHz after annealing at 773 K (Figure 8b). The minimum core loss is also obtained for the SMC annealed at 773 K (Figure 8c).

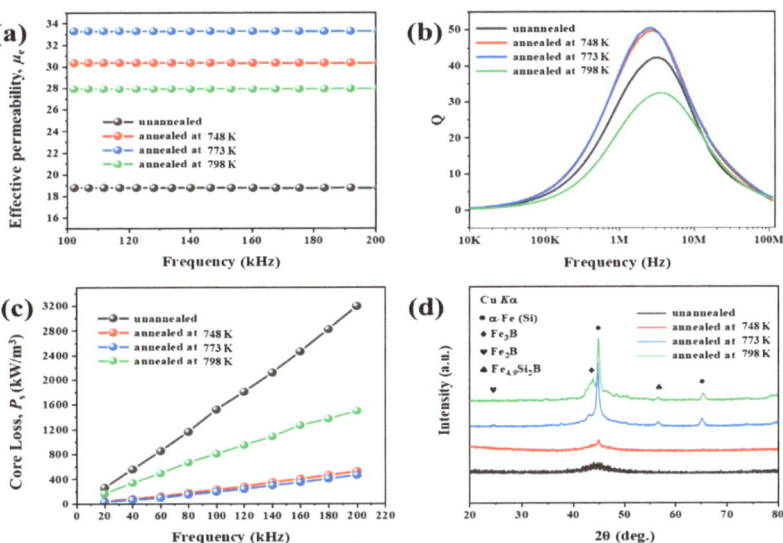

Figure 8. Frequency dependence of the (**a**) μ_e, (**b**) Q, (**c**) P_s, (**d**) XRD for SMCs after annealing at different temperatures.

To illustrate the reasons for the deterioration of magnetic properties after annealing above 798 K, the magnetic powders were tested for XRD, as shown in Figure 7. For the original powders without annealing, only one broad diffraction peak can be observed, indicating a typical amorphous structure [37]. For the powders annealed at the temperatures below 748 K, a slightly intense α-Fe (Si) phase begins to appear. The crystallization occurred after the sample annealed at 773 K, with the precipitation of α-Fe(Si) and Fe$_3$B phases. When the annealed temperature rises to 798 K, the intensity of Fe$_3$B phase increases. Compared with the soft magnetic phase of α-Fe(Si). the Fe$_3$B phase has a higher anisotropic field, resulting in decreased soft magnetic properties of the SMCs [38]. The average grain size D of α-Fe(Si) annealed at 773 K is approximately 18 nm calculated by the Scherrer formula. These nano-sized grains are randomly dispersed in the residual amorphous matrix. When D < L$_{ex}$ (L$_{ex}$ is value of ferromagnetic exchange length, about 30~40 nm for α-Fe), with the reduction of grain size, the effective anisotropy shows a sharp decrease and the impact is negligible. The magnetic domain structure exhibits a vortex structure determined by the exchange coupling energy and demagnetization energy jointly, the coercive force and residual magnetism are close to zero. Therefore, the magnetic powders after annealing at 773 K have excellent soft magnetic properties [39,40].

3.4. DC-Bias Properties

The presence of DC current or voltage components in an AC power system is called DC bias. The DC-bias property is important for SMC since almost all of the powder cores are used in a DC-bias field. Figure 9 shows the percentage of permeability as a function of the DC-bias field for SMC cores, which were made from the powders coated with different H$_3$PO$_4$ concentrations followed by 2 wt.% sodium silicate solution and annealed at 773 K. The permeability decreases with the increase of the bias field for all cores, since the SMCs approach to magnetic saturation at high DC field. In addition, with the increase of H_3PO_4 concentrations from 0 to 0.75 wt.%, the DC-bias performance of the core, defined by the percentage of reduced permeability, increased from 75.3 to 88.2% at 4000 A/m. The results indicate that the H_3PO_4 and sodium silicate double-layer coating can significantly improve the DC bias performance of SMCs. In particular, the SMCs prepared with H_3PO_4 concentrations of 0.5 and 0.75 wt.% exhibit relatively superior DC-bias performances (over 80%). The reason is that the voids and gaps in the powder cores can pin the domain wall in the magnetizing process and prevent the propagation of domain movement between particles, then suppressing the decrease of permeability, which is beneficial in achieving a stable percent permeability [41].

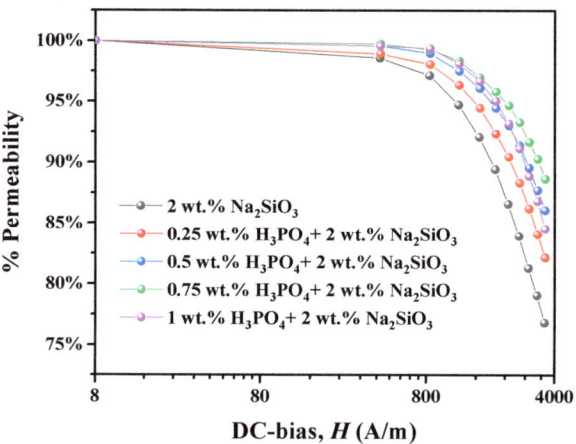

Figure 9. The variation of the permeability with the DC-bias field for SMCs after coating with different H_3PO_4 concentrations.

3.5. Radial Crushing Properties

The SMCs based on powders coated by different H_3PO_4 concentrations and 2 wt.% sodium silicate-aqueous solution followed by annealed at 773 K is employed for radial crush testing. The test method is shown in Figure 10a, where the radial crush strength of a ring shape specimen is determined by applying radial pressure. Figure 10b shows the radial crush strength of various SMCs. The strength first increases then decreases with the increasing concentration. The $Si(OH)_4$ produced by hydrolysis of sodium silicate, through dehydration and condensation between -OH can form networked SiO_2 between the magnetic powders, which can increase the binding force between the magnetic powders. When the powders coated by 2 wt.% sodium silicate-aqueous solution alone, the binding force between the powder is poor due to the weak binding force between SiO_2 and powders, which is manifested by the low radial crush strength of the SMCs. As the H_3PO_4 concentration increases, the radial crush strength increases since the rough surface of the phosphate can increase the binding force with SiO_2, while the binding force between the powders also increases. However, when the H_3PO_4 concentration exceeds 0.5 wt.%, the radial crush strength of the sample decreases, since part of the SiO_2 forms on sheets detached from phosphate layer.

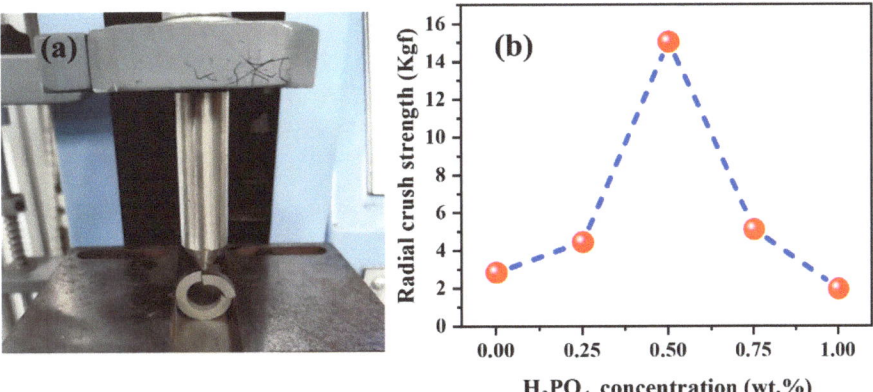

Figure 10. (a) The radial crush test method and (b) the radial crush strength of SMCs based on powders prepared by different H_3PO_4 concentration followed by 2 wt.% sodium silicate-aqueous solution after annealing.

4. Conclusions

The core-shell structured amorphous FeSiBCr@phosphate/SiO_2 powders were prepared by H_3PO_4 and sodium silicate treatments for fabricating high performance SMCs. The processes of phosphating, sodium silicate coating and annealing treatment have been investigated in detail. A complete and uniform phosphate/SiO_2 doubled layer has been prepared on the amorphous powder by 0.5 wt.% H_3PO_4 followed by 2 wt.% sodium silicate treatments. When the H_3PO_4 concentration exceeds to 0.5 wt.%, the phosphate sheets were detached from phosphate layer, leading to poor powder coating efficiency and poor performance of SMC. After annealing at 773 K, the SMCs show improved permeability and greatly reduced core loss due to the effectively released residual stress. However, when the annealing temperature is 798 K, the soft magnetic properties including permeability and core loss were deteriorated due to the increasing precipitation of Fe_3B phase. The excellent magnetic properties with permeability $\mu_e = 33.5$ and core loss $P_s = 368$ kW/m^3 at 50 mT and 200 kHz have been achieved in the SMC prepared by optimized process. i.e., 0.5 wt.% H_3PO_4 -ethanol solution and 2 wt.% sodium silicate-aqueous solution treatment followed by 773 K annealing, which promises great potential applications of electronic components at medium and high frequencies.

Author Contributions: Conceptualization, P.L. and C.W.; methodology, P.L. and H.Y. (Han Yuan); software, P.L.; validation, P.L., H.Y. (Hongya Yu) and Z.L.; formal analysis, P.L.; investigation, P.L. and C.W.; resources, H.Y. (Han Yuan); data curation, P.L.; writing—original draft preparation, P.L.; writing—review and editing, H.Y. (Hongya Yu), Z.L., and P.L.; visualization, C.W.; supervision, Z.L., Y.W., L.Y. and W.W.; project administration, H.Y. (Hongya Yu), Z.L.; funding acquisition, Y.W., L.Y. and W.W.; All authors have read and agreed to the published version of the manuscript.

Funding: This research was funded by Southwest Institute of Applied Magnetism, grant number 2022SK-008 and Guangdong Provincial Natural Science Foundation of China, 2021A1515010642, grant number.

Data Availability Statement: Not applicable.

Conflicts of Interest: The authors declare no conflict of interest.

References

1. Shokrollahi, H.; Janghorban, K. Soft magnetic composite materials (SMCs). *J. Mater. Process. Technol.* **2007**, *189*, 1–12. [CrossRef]
2. Sunday, K.J.; Taheri, M.L. Soft magnetic composites: Recent advancements in the technology. *Met. Powder Rep.* **2017**, *72*, 425–429. [CrossRef]

3. Périgo, E.A.; Weidenfeller, B.; Kollár, P.; Füzer, J. Past, present, and future of soft magnetic composites. *Appl. Phys. Rev.* **2018**, *5*, 031301. [CrossRef]
4. Silveyra, J.M.; Ferrara, E.; Huber, D.L.; Monson, T.C. Soft magnetic materials for a sustainable and electrified world. *Science* **2018**, *362*, 418–427. [CrossRef]
5. Makino, A.; Kubota, T.; Makabe, M.; Chang, C.T.; Inoue, A. FeSiBP metallic glasses with high glass-forming ability and excellent magnetic properties. *Mater. Sci. Eng. B* **2008**, *148*, 166–170. [CrossRef]
6. Wang, J.; Li, R.; Hua, N.; Huang, L.; Zhang, T. Ternary Fe–P–C bulk metallic glass with good soft-magnetic and mechanical properties. *Scr. Mater.* **2011**, *65*, 536–539. [CrossRef]
7. Li, Z.; Wang, A.; Chang, C.; Wang, Y.; Dong, B.; Zhou, S. Synthesis of FeSiBPNbCu nanocrystalline soft-magnetic alloys with high saturation magnetization. *J. Alloys Compd.* **2014**, *611*, 197–201. [CrossRef]
8. Sun, H.; Wang, C.; Wang, J.; Yu, M.; Guo, Z. Fe-based amorphous powder cores with low core loss and high permeability fabricated using the core-shell structured magnetic flaky powders. *J. Magn. Magn. Mater.* **2020**, *502*, 166548. [CrossRef]
9. Wang, C.; Guo, Z.; Wang, J.; Sun, H.; Chen, D.; Chen, W.; Liu, X. Industry-oriented Fe-based amorphous soft magnetic composites with SiO_2-coated layer by one-pot high-efficient synthesis method. *J. Magn. Magn. Mater.* **2020**, *509*, 166924. [CrossRef]
10. Zhou, B.; Dong, Y.; Chi, Q.; Zhang, Y.; Chang, L.; Gong, M.; Huang, J.; Pan, Y.; Wang, X. Fe-based amorphous soft magnetic composites with SiO_2 insulation coatings: A study on coatings thickness, microstructure and magnetic properties. *Ceram. Int.* **2020**, *46*, 13449–13459. [CrossRef]
11. Zhang, Y.; Zhou, T.d. Structure and electromagnetic properties of FeSiAl particles coated by MgO. *J. Magn. Magn. Mater.* **2017**, *426*, 680–684. [CrossRef]
12. Zhou, B.; Dong, Y.; Liu, L.; Chi, Q.; Zhang, Y.; Chang, L.; Bi, F.; Wang, X. The core-shell structured Fe-based amorphous magnetic powder cores with excellent magnetic properties. *Adv. Powder Technol.* **2019**, *30*, 1504–1512. [CrossRef]
13. Peng, Y.; Yi, Y.; Li, L.; Yi, J.; Nie, J.; Bao, C. Iron-based soft magnetic composites with Al_2O_3 insulation coating produced using sol-gel method. *Mater. Des.* **2016**, *109*, 390–395. [CrossRef]
14. Geng, K.; Xie, Y.; Yan, L.; Yan, B. Fe-Si/ZrO_2 composites with core-shell structure and excellent magnetic properties prepared by mechanical milling and spark plasma sintering. *J. Alloys Compd.* **2017**, *718*, 53–62. [CrossRef]
15. Wang, Z.; Liu, X.; Kan, X.; Zhu, R.; Yang, W.; Wu, Q.; Zhou, S. Preparation and characterization of flaky FeSiAl composite magnetic powder core coated with MnZn ferrite. *Curr. Appl. Phys.* **2019**, *19*, 924–927. [CrossRef]
16. Birčáková, Z.; Onderko, F.; Dobák, S.; Kollár, P.; Füzer, J.; Bureš, R.; Fáberová, M.; Weidenfeller, B.; Bednarčík, J.; Jakubčin, M.; et al. Eco-friendly soft magnetic composites of iron coated by sintered ferrite via mechanofusion. *J. Magn. Magn. Mater.* **2022**, *543*, 168627. [CrossRef]
17. Liu, D.; Wu, C.; Yan, M.; Wang, J. Correlating the microstructure, growth mechanism and magnetic properties of FeSiAl soft magnetic composites fabricated via HNO_3 oxidation. *Acta Mater.* **2018**, *146*, 294–303. [CrossRef]
18. Wei, H.; Yu, H.; Feng, Y.; Wang, Y.; He, J.; Liu, Z. High permeability and low core loss nanocrystalline soft magnetic composites based on FeSiBNbCu@Fe_3O_4 powders prepared by HNO_3 oxidation. *Mater. Chem. Phys.* **2021**, *263*, 124427. [CrossRef]
19. Yu, H.; Zhou, S.; Zhang, G.; Dong, B.; Meng, L.; Li, Z.; Dong, Y.; Cao, X. The phosphating effect on the properties of FeSiCr alloy powder. *J. Magn. Magn. Mater.* **2022**, *552*, 168741. [CrossRef]
20. Chen, Z.; Liu, X.; Kan, X.; Wang, Z.; Zhu, R.; Yang, W.; Wu, Q.; Shezad, M. Phosphate coatings evolution study and effects of ultrasonic on soft magnetic properties of FeSiAl by aqueous phosphoric acid solution passivation. *J. Alloys Compd.* **2019**, *783*, 434–440. [CrossRef]
21. Ding, W.; Jiang, L.; Li, B.; Chen, G.; Tian, S.; Wu, G. Microstructure and Magnetic Properties of Soft Magnetic Composites with Silicate Glass Insulation Layers. *J. Supercond. Novel. Magn.* **2013**, *27*, 239–245. [CrossRef]
22. Wang, D.L.; Ding, H.P.; Ma, Y.F.; Gong, P.; Wang, X.Y. Research progress on corrosion resistance of metallic glasses. *J. Chin. Soc. Corros. Prot.* **2021**, *41*, 277–288.
23. Chen, S.F.; Chang, H.Y.; Wang, S.J.; Chen, S.H.; Chen, C.C. Enhanced electromagnetic properties of Fe–Cr–Si alloy powders by sodium silicate treatment. *J. Alloys Compd.* **2015**, *637*, 30–35. [CrossRef]
24. Hsiang, H.I.; Wang, S.K.; Chen, C.C. Electromagnetic properties of FeSiCr alloy powders modified with amorphous SiO_2. *J. Magn. Magn. Mater.* **2020**, *514*, 167151. [CrossRef]
25. Sun, X.; Zhan, J.; Peng, Z. Prepration and characterization of nano-titania surface modified with silica. *J. Chin. Ceram. Soc.* **2007**, *9*, 1174–1177.
26. Wang, J.; Liu, X.; Li, L.; Mao, X. Performance improvement of Fe−6.5Si soft magnetic composites with hybrid phosphate-silica insulation coatings. *J. Cent. South Univ.* **2021**, *28*, 1266–1278. [CrossRef]
27. Liu, D.; Wu, C.; Yan, M. Investigation on sol–gel $Al2O3$ and hybrid phosphate-alumina insulation coatings for FeSiAl soft magnetic composites. *J. Mater. Sci.* **2015**, *50*, 6559–6566. [CrossRef]
28. Li, K.; Cheng, D.; Yu, H.; Liu, Z. Process optimization and magnetic properties of soft magnetic composite cores based on phosphated and mixed resin coated Fe powders. *J. Magn. Magn. Mater.* **2020**, *501*, 166455. [CrossRef]
29. Han, H.; Duan, D.; Yuan, P. Binders and Bonding Mechanism for RHF Briquette Made from Blast Furnace Dust. *ISIJ Int.* **2014**, *54*, 1781–1789. [CrossRef]
30. Luo, Z.; Fan, X.; Hu, W.; Luo, F.; Li, G.; Li, Y.; Liu, X.; Wang, J. Controllable SiO_2 insulating layer and magnetic properties for intergranular insulating Fe-6.5wt.%Si/SiO_2 composites. *Adv. Powder Technol.* **2019**, *30*, 538–543. [CrossRef]

31. Chang, L.; Xie, L.; Liu, M.; Li, Q.; Dong, Y.; Chang, C.; Wang, X.M.; Inoue, A. Novel Fe-based nanocrystalline powder cores with excellent magnetic properties produced using gas-atomized powder. *J. Magn. Magn. Mater.* **2018**, *452*, 442–456. [CrossRef]
32. Wu, Z.Y.; Jiang, Z.; Fan, X.A.; Zhou, L.J.; Wang, W.L.; Xu, K. Facile synthesis of Fe-6.5wt%Si/SiO2 soft magnetic composites as an efficient soft magnetic composite material at medium and high frequencies. *J. Alloys Compd.* **2018**, *742*, 90–98. [CrossRef]
33. Chang, C.; Guo, J.; Li, Q.; Zhou, S.; Liu, M.; Dong, Y. Improvement of soft magnetic properties of FeSiBPNb amorphous powder cores by addition of FeSi powder. *J. Alloys Compd.* **2019**, *788*, 1177–1181. [CrossRef]
34. Hsiang, H.I.; Fan, L.F.; Hung, J.J. Phosphoric acid addition effect on the microstructure and magnetic properties of iron-based soft magnetic composites. *J. Magn. Magn. Mater.* **2018**, *447*, 1–8. [CrossRef]
35. Li, Z.; Chen, L. Static and dynamic magnetic properties of Co_2Z barium ferrite nanoparticle composites. *J. Mater. Sci.* **2005**, *40*, 719–723. [CrossRef]
36. Shokrollahi, H.; Janghorban, K. Effect of warm compaction on the magnetic and electrical properties of Fe-based soft magnetic composites. *J. Magn. Magn. Mater.* **2007**, *313*, 182–186. [CrossRef]
37. Sugimura, K.; Yabu, N.; Sonehara, M.; Sato, T. Novel Method for Making Surface Insulation Layer on Fe-Based Amorphous Alloy Powder by Surface-Modification Using Two-Step Acid Solution Processing. *IEEE Trans. Magn.* **2018**, *54*, 1–5. [CrossRef]
38. Zhou, B.; Chi, Q.; Dong, Y.; Liu, L.; Zhang, Y.; Chang, L.; Pan, Y.; He, A.; Li, J.; Wang, X. Effects of annealing on the magnetic properties of Fe-based amorphous powder cores with inorganic-organic hybrid insulating layer. *J. Magn. Magn. Mater.* **2020**, *494*, 165827. [CrossRef]
39. Li, T.; Dong, Y.; Liu, L.; Liu, M.; Shi, X.; Dong, X.; Rong, Q. Novel Fe-based nanocrystalline powder cores with high performance prepared by using industrial materials. *Intermetallics* **2018**, *102*, 101–105. [CrossRef]
40. Narayanan, T. Surface pretreatment by phosphate conversion coatings—A review. *Rev. Adv. Mater. Sci.* **2005**, *9*, 130–177.
41. Dong, Y.; Li, Z.; Liu, M.; Chang, C.; Li, F.; Wang, X.M. The effects of field annealing on the magnetic properties of FeSiB amorphous powder cores. *Mater. Res. Bull.* **2017**, *96*, 160–163. [CrossRef]

Disclaimer/Publisher's Note: The statements, opinions and data contained in all publications are solely those of the individual author(s) and contributor(s) and not of MDPI and/or the editor(s). MDPI and/or the editor(s) disclaim responsibility for any injury to people or property resulting from any ideas, methods, instructions or products referred to in the content.

Article

Effects of La on Thermal Stability, Phase Formation and Magnetic Properties of Fe–Co–Ni–Si–B–La High Entropy Alloys

Jiaming Li, Jianliang Zuo and Hongya Yu *

School of Materials Science and Engineering, South China University of Technology, Guangzhou 510640, China; johnlee1215@163.com (J.L.); zuojianliang@126.com (J.Z.)
* Correspondence: yuhongya@scut.edu.cn

Abstract: The microstructure, phase formation, thermal stability and soft magnetic properties of melt-spun high entropy alloys (HEAs) $Fe_{27}Co_{27}Ni_{27}Si_{10-x}B_9La_x$ with various La substitutions for Si (x = 0, 0.2, 0.4, 0.6, 0.8, and 1) were investigated in this work. The $Fe_{27}Co_{27}Ni_{27}Si_{10-x}B_9La_{0.6}$ alloy shows superior soft magnetic properties with low coercivity H_c of ~7.1 A/m and high saturation magnetization B_s of 1.07 T. The content of La has an important effect on the primary crystallization temperature (T_{x1}) and the secondary crystallization temperature (T_{x2}) of the alloys. After annealing at relatively low temperature, the saturation magnetization of the alloy increases and the microstructure with a small amount of body-centered cubic (BCC) phase embedded in amorphous matrix is observed. Increasing the annealing temperature reduces the magnetization due to the transformation of BCC phase into face-centered cubic (FCC) phase.

Keywords: magnetic materials; high entropy alloys; thermal stability; phase transformation

Citation: Li, J.; Zuo, J.; Yu, H. Effects of La on Thermal Stability, Phase Formation and Magnetic Properties of Fe–Co–Ni–Si–B–La High Entropy Alloys. *Metals* **2021**, *11*, 1907. https://doi.org/10.3390/met11121907

Academic Editor: Jiro Kitagawa

Received: 5 October 2021
Accepted: 16 November 2021
Published: 26 November 2021

Publisher's Note: MDPI stays neutral with regard to jurisdictional claims in published maps and institutional affiliations.

Copyright: © 2021 by the authors. Licensee MDPI, Basel, Switzerland. This article is an open access article distributed under the terms and conditions of the Creative Commons Attribution (CC BY) license (https://creativecommons.org/licenses/by/4.0/).

1. Introduction

High entropy alloys (HEAs), defined as the alloys consisting of at least five principal elements without obvious base element, have been proposed by Cantor et al. [1] and Yeh et al. [2] in 2004, independently. Up to now, there are two commonly used definitions of HEAs. One is the composition-based concept, i.e., the alloys composed of five or more principal elements in equal or near equal molar ratio between 5 atom percent (at.%) and 35 atom percent (at.%). The other definition is based on total configurational molar entropy (S_{mix}). The alloys with S_{mix} < 1 R, 1 R < S_{mix} < 1.5 R, and S_{mix} >1.5 R, where R is the gas constant, are defined as low entropy alloys, medium entropy alloys, and high entropy alloys, respectively [3,4]. As a new type of alloys with unique properties of high strength, hardness, corrosion resistance, abrasion resistance and high fatigue resistance, HEAs have received extensive attention. Instead of forming a complex crystal structure, HEAs usually tend to form a solid solution with a face-centered cubic (FCC) or body-centered cubic (BCC) structure, or a mixture thereof [5,6], although a hexagonal close-packed (HCP) structure may be found in a few of HEAs [7].

Studying the compositions, microstructure and their fundamental properties to establish a fundamental database is currently the most essential work for HEAs [8,9]. Up to now, a series of HEAs have been prepared, including Fe-based, Co-based [10], Fe–Co–Ni-based [11] and rare earth-based high-entropy alloys [12,13]. However, most of the previous work focused on their mechanical properties and microstructure [9,14], and their physical properties have not been fully investigated. As Fe, Co and Ni are common constituent elements used in HEAs [15], it is very interesting to explore the magnetic properties of the HEAs. As we know, the soft magnetic materials are developing towards low coercivity (H_c) and high magnetization (M_s), which are essential for promoting the energy conservation efficiency and miniaturization of the electromagnetic device. However, some existing reports on the high-entropy soft magnetic alloys indicate that the saturation induction B_s

of the HEAs is still low, typically less than 1 T and their crystallization temperature (T_x) is also less than 670 K [16–19], which are both less than what we expected.

On the other hand, the rare earth elements (RE) have been frequently employed in the soft magnetic alloys [20,21], and the results showed that the addition of RE elements such as Gd and Tb can increase the curie temperature [20], and modify the crystallization temperature of the alloy. The addition of RE can also decrease the magnetic permeability. However, the influence of trace rare earth elements on HEAs has rarely been studied. In this work, La is selected to substitute Si for improving the performance of Fe–Co–Ni–Si–B HEAs. La exhibits low solubility with Fe, Co, and Ni elements, and it may play a role of micro-alloying. The thermodynamic properties, glass-forming ability (GFA) and magnetic properties of Fe–Co–Ni–Si–B–La HEAs are studied in detail.

2. Experimental

The alloy ingots of $Fe_{27}Co_{27}Ni_{27}Si_{10-x}B_9La_x$ with x = 0, 0.2, 0.4, 0.6, 0.8, and 1 (atomic ratio), denoted as La_0, $La_{0.2}$, $La_{0.4}$, $La_{0.6}$, $La_{0.8}$, and La_1, respectively, were prepared by arc-melting pure Fe (99.5 wt.%), Co (99.9 wt.%), Ni (99.96 wt.%), La (99.9 wt.%) metals, FeB (with Fe 83.78 wt.% and B 16.22 wt.%) and Si (99.99 wt.%) crystals under argon atmosphere. The ingots were melted 5 times to ensure chemical homogeneity. The ribbons with width of ~1.2 mm and thickness of ~0.025 mm were prepared by single-roller melt spinning method with the wheel speeds of 45 m/s. The phase structures of the alloys were characterized by X-ray diffraction (XRD) with Cu Kα radiation. Thermal stability was studied by differential scanning calorimetry (DSC) at a heating rate of 10 K/min and under argon atmosphere. The saturation magnetization (B_s) of ribbons were measured under an applied field of 250 kA/m with a vibrating sample magnetometer (VSM). The coercive force (H_c) was measured with a MATS-2010SD hysteresis curve (DC) test system using ribbons about 50 mm in length.

3. Results and Discussion

It is reported that HEAs trend to form simple fcc and/or bcc solid solution structure or metallic glass. Figure 1 shows the XRD pattern of arc melt $Fe_{27}Co_{27}Ni_{27}Si_{10-x}B_9La_x$ (x = 0, 0.6, and 1) ingots. In all the ingots, the fcc phase, $(FeCoNi)_2B$ and $Ni_{31}Si_{12}$ phases were detected [18,19].

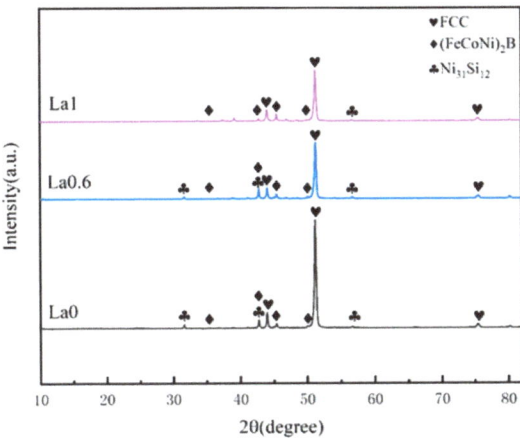

Figure 1. The XRD pattern of as-cast $Fe_{27}Co_{27}Ni_{27}Si_{10-x}B_9La_x$ (x = 0, 0.6, 1) ingots.

According to prior research [22], there are many phases in high entropy alloys, including solid solutions, intermetallic compounds, and amorphous phases. The phase evolution in HEAs can be predicted mainly by three parameters, namely atomic size difference

(Δ), mixing enthalpy (ΔH_{mix}), mixing entropy (ΔS_{mix}) and valence electron concentration (VEC) [23,24]. The Δ, ΔH_{mix}, ΔS_{mix}, and VEC are defined as:

$$\delta = 100\sqrt{\sum_{i=1}^{N} c_i(1 - \frac{r_i}{\bar{r}})^2} \quad (1)$$

$$\Delta H_{mix} = \sum_{i=1, i \neq j}^{N} \Omega_{ij} c_i c_j \quad (2)$$

$$\Delta S_{mix} = -R \sum_{i=1}^{N} c_i \ln c_i \quad (3)$$

$$VEC = \sum_{i=1}^{N} c_i (VEC)_i \quad (4)$$

where N is the number of the components in HEAs, R is gas constant, c_i is the atomic fraction of i-th component, and $\bar{r}$ is the average atomic radius. r_i is the atomic radius, which can be obtained from References [15,24]. VEC, ΔH_{mix}, and ΔS_{mix} between atomic pairs also be obtained in References [15,24]. The values of Δ, ΔH_{mix}, ΔS_{mix}, and VEC for $Fe_{27}Co_{27}Ni_{27}Si_{10-x}B_9La_x$ alloys are summarized in Table 1. It is clear that all VEC values are near 8.0. Guo et al. [25] pointed out that fcc phase forms in the alloy with $VEC \geq 8.0$, bcc phase forms at $VEC \leq 6.87$, and a mixture of fcc and bcc phases at $6.87 \leq VEC \leq 8.0$. Hence, these $Fe_{27}Co_{27}Ni_{27}Si_{10-x}B_9La_x$ alloys trend to form fcc solid solution and intermetallic compounds.

Table 1. The atomic radius difference (Δ), valence electron concentration (VEC), mixing enthalpy (ΔH_{mix}), mixing entropy (ΔS_{mix}) and structure of the $Fe_{27}Co_{27}Ni_{27}Si_{10-x}B_9La_x$ (x = 0, 0.6, 1) alloy systems (atomic percent).

Sample	Δ (%)	ΔS (kJ/mol)	ΔH (kJ/mol)	VEC	Structure
$Fe_{27}Co_{27}Ni_{27}Si_{10}B_9$	10.2	12.615	−20.42	7.96	FCC + IM
$Fe_{27}Co_{27}Ni_{27}Si_9B_9La_{0.6}$	11.06	12.722	−19.31	7.954	FCC + IM
$Fe_{27}Co_{27}Ni_{27}Si_9B_9La_1$	11.58	12.804	−18.66	7.95	FCC + IM

Figure 2a shows the XRD patterns of as-spun $Fe_{27}Co_{27}Ni_{27}Si_{10-x}B_9La_x$ alloys. Only a broad diffraction peak appears at near 45° without any detectable crystalline peaks for all alloys, indicating fully amorphous structure. Figure 2b shows the DSC curves of the as-spun $Fe_{27}Co_{27}Ni_{27}Si_{10-x}B_9La_x$ ribbons. All curves have two distinct exothermic peaks and one endothermic peak, giving two-stage crystallization and melting processes. The glass transition temperature (T_g), phase transition temperature (T_p) [16], liquidus temperature (T_l), primary crystallization temperature (T_{x1}), and secondary crystallization temperature (T_{x2}) are marked by arrows. As shown in Figure 2b, The T_g of amorphous ribbons range from 642 to 694 K. The T_{x1} and T_{x2} for the alloys with different La contents are in the region of 707–743 K and 802–839 K, respectively. The T_{x1} initially increases from 722 to 743 K with increasing La content from 0 to 0.2 at.%, and then decreases to 707 K with increasing La to 1 at.%. The largest T_{x1} of 743 K is obtained in the alloy with 0.2 at.% La substitution. Similarly, T_{x2} increase from 802 to 839 K with further increase of La. At 1 at.% La substitution, T_{x2} reaches the largest value of about 839 K. The value of ΔT_x (= $T_x - T_g$) of these alloys are in the region of 43–65 K, and it becomes large as x increases up to 1, which indicates that less than 1 at.% La substitution is beneficial to forming amorphous structure and hindering crystallization process [26]. The large ΔT_x up to 65 K for $Fe_{27}Co_{27}Ni_{27}Si_9B_9La_1$ alloy shows good thermal stability of the supercooled liquid. In addition, T_l of alloys increases from 1318 to 1439 K as x increase from 0 to 0.6 at.%, then decreases to 1324 K with x increases to 1 at.%. The T_g, T_x, T_l, ΔT_x, reduced glass transition temperature T_{rg} ($= T_g/T_l$) [27], and S ($= \Delta T_x/(T_l - T_g)$) [28] are listed in Table 2. The S values and T_{rg} values exhibit good correlation with ΔT_x, and the largest S value of 0.096 is obtained at x = 1.

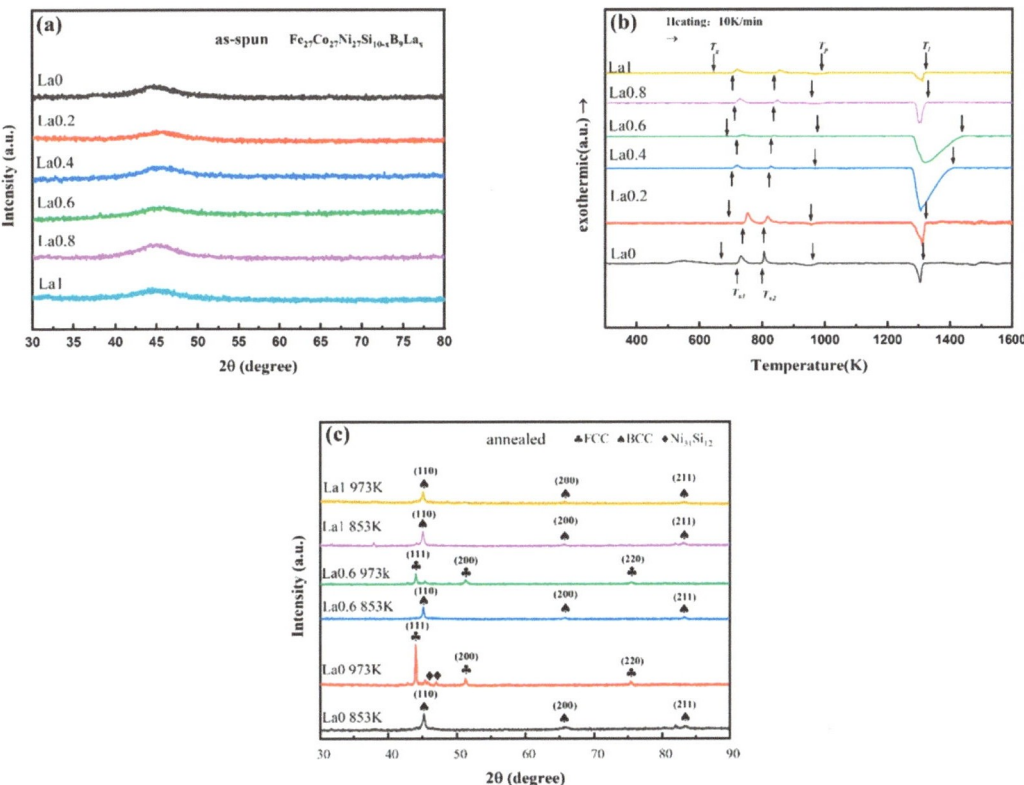

Figure 2. (a) XRD patterns of as-spun $Fe_{27}Co_{27}Ni_{27}Si_{10-x}B_9La_x$ (x = 0 to 1) alloys, (b) DSC curve of as-spun $Fe_{27}Co_{27}Ni_{27}Si_{10-x}B_9La_x$ alloys, and (c) XRD patterns of amorphous ribbons annealed for 5 min at 853 and 973 K.

Table 2. Thermal parameters of $Fe_{27}Co_{27}Ni_{27}Si_{10-x}B_9La_x$ amorphous ribbons.

Composition (at.%)	T_g (K)	T_{x1} (K)	T_{x2} (K)	ΔT_x (K)	T_l (K)	T_{rg}	S
$Fe_{27}Co_{27}Ni_{27}Si_{10}B_9$	667	722	802	55	1318	0.506	0.084
$Fe_{27}Co_{27}Ni_{27}Si_{9.8}B_9La_{0.2}$	694	743	806	49	1324	0.524	0.078
$Fe_{27}Co_{27}Ni_{27}Si_{9.8}B_9La_{0.4}$	-	710	823	-	1407	-	-
$Fe_{27}Co_{27}Ni_{27}Si_{9.8}B_9La_{0.6}$	680	723	830	43	1439	0.473	0.057
$Fe_{27}Co_{27}Ni_{27}Si_{9.8}B_9La_{0.8}$	-	713	832	-	1327	-	-
$Fe_{27}Co_{27}Ni_{27}Si_{9.8}B_9La_1$	642	707	839	65	1321	0.486	0.096

To further study the crystallization behavior of $Fe_{27}Co_{27}Ni_{27}Si_{10-x}B_9La_x$ ribbons with different La contents, the $Fe_{27}Co_{27}Ni_{27}Si_{10-x}B_9La_x$ (x = 0, 0.6, 1) ribbons were annealed at different temperatures. Figure 2c shows the XRD patterns of $Fe_{27}Co_{27}Ni_{27}Si_{10-x}B_9La_x$ (x = 0, 0.6, 1) alloys after annealing at 853 and 973 K (above T_p) for 5 min. After annealing at 853 K, between T_{x1} and T_{x2}, a bcc phase precipitates in the amorphous matrix. With the annealing temperature increased to 973 K, the bcc phase disappeared and fcc crystals formed. The transformation of bcc phase to fcc phase can also be observed in Fe–Co–Ni–Si–B HEAs at high temperature [29]. Combined with the DSC analysis results, the first exothermic peak is due to the precipitation of bcc phase and the second peak originates from bcc phase and $Ni_{31}Si_{12}$ phases, and T_p represents the transformation of bcc phase to fcc phase. For the alloy without La-substitution, after annealing at 973 K, a small amount of intermetallic compounds was detected, indexed as $Ni_{31}Si_{12}$ phase. Previous research has demonstrated that the Ni element is easy to segregate from Fe-rich bcc phases, resulting in the formation of fcc phase, and the over-saturated Si in Ni may form the Ni-Si intermetallic compounds [29,30]. However, after La addition, no $Ni_{31}Si_{12}$ phase was detected in annealed samples. Previous study [29] also showed that in Fe–Co–Ni–Si–B HEAs, $Ni_{31}Si_{12}$ phase could exist in high temperature. Thus, in the present alloys, the addition of La can suppress the formation of $Ni_{31}Si_{12}$ phase. Based on above discussion, the phase transition in $Fe_{27}Co_{27}Ni_{27}Si_{10-x}B_9La_x$ amorphous alloys after annealing occurs through the process of amorphous → amorphous' + bcc phase + $Ni_{31}Si_{12}$ → fcc phase.

Figure 3a shows the magnetic hysteresis loops (M–H curves) of as-spun $Fe_{27}Co_{27}Ni_{27}Si_{10-x}B_9La_x$ ($x = 0$ to 1) alloys. All alloys show soft magnetic behavior. The saturation magnetization M_s of these alloys increases from 0.86 T to 1.01 T as x increases from 0 to 0.4, and then decreases to 0.88 T with x increasing to 1. Figure 3b shows the M–H curves of the $Fe_{27}Co_{27}Ni_{27}Si_{10-x}B_9La_x$ ($x = 0, 0.6, 1$) alloys after annealing at 573 K (below glass transition temperature) and 703 K (below crystallization temperature) for 5 min. The saturation magnetization M_s increases with the increasing annealing temperature. The M_s values of the alloys with x = 0, 0.6 and 1 annealed at 703 K is about 0.96 T, 0.99 T and 0.97 T, respectively. The coercivity H_c values were measured as 10.3, 7.1, and 8.5 A/m at 573 K, respectively. As the heat treatment temperature increased to 703 K, the coercivity values were obtained as 18.4, 22.6 and 12.6 A/m for $Fe_{27}Co_{27}Ni_{27}Si_{10-x}B_9La_x$ ($x = 0, 0.6$, and 1) alloys.

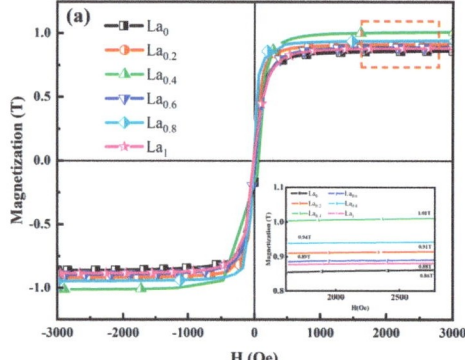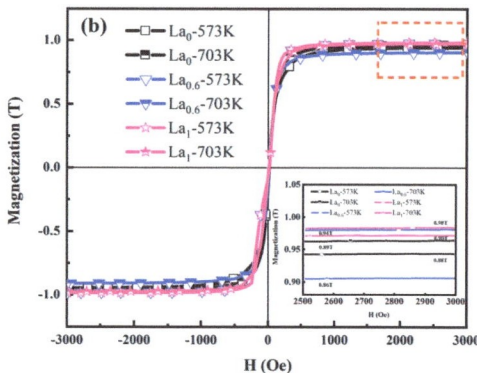

Figure 3. (a) The M–H curves of as-spun $Fe_{27}Co_{27}Ni_{27}Si_{10-x}B_9La_x$ ($x = 0$ to 1) alloys, and (b) the M–H curves of $Fe_{27}Co_{27}Ni_{27}Si_{10-x}B_9La_x$ ($x = 0, 0.6, 1$) alloys after heat treatment at different temperatures for 5 min.

As the annealing temperature rises above the crystallization temperature, the solid solution phase or other phases precipitate in the alloy. It is important to confirm the influence of the precipitation of the solid solution phase on the soft magnetic properties of HEAs. The magnetization curves of the annealing $Fe_{27}Co_{27}Ni_{27}Si_{10-x}B_9La_x$ ($x = 0, 0.6$, and 1) ribbons are shown in Figure 4. After annealing at 853 K, the saturation magnetization of the HEAs ribbons is increased. The M_s values of the alloys with $x = 0, 0.6$ and 1 annealed at 853 K is about 1.05 T, 1.06 T and 1.07 T, respectively. With increasing annealing temperature to 973 K, M_s and H_c decrease simultaneously. The M_s values of the alloys with x = 0, 0.6 and 1 is about 0.93 T, 0.89 T and 1.0 T, respectively. After annealing at higher temperature, the values of M_s increased, but the coercivity was deteriorated dramatically. As shown in Figure 5, the coercivity H_c, after crystallization annealing is greatly increased. This phenomenon may be related to the fine grains precipitated in the amorphous matrix. Small crystal grains hinder the movement of magnetic domains and play a pinning role. According to the current experimental data, the addition of La element can increase the recrystallization temperature, so an appropriate amount of La can reduce the effect of heat treatment on the reduction of saturation magnetization (sample La_1 have highest B_s after annealing at 973 K). At the same time, the coercivity of the alloys with La element after annealing is relatively small.

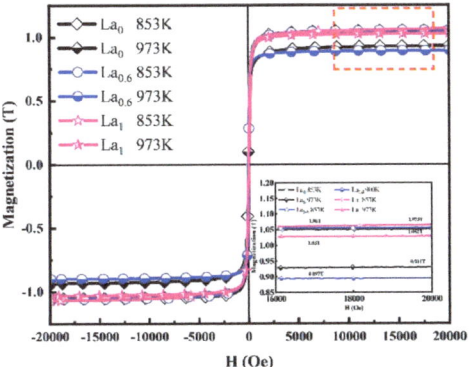

Figure 4. The M–H curves of $Fe_{27}Co_{27}Ni_{27}Si_{10-x}B_9La_x$ ($x = 0, 0.6, 1$) alloys annealed above the crystallization temperature.

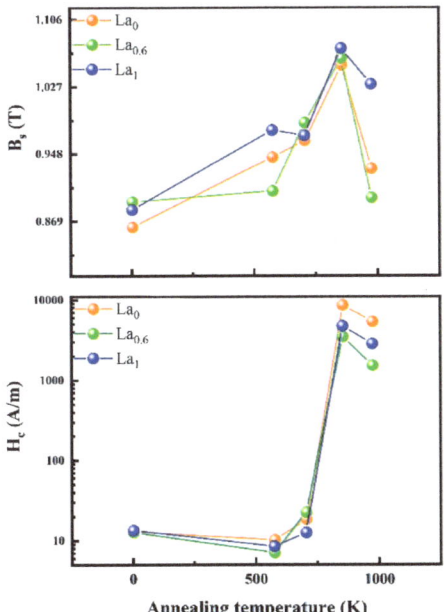

Figure 5. Changes in saturation magnetization (B_s), coercivity (H_c) with different annealing temperature for $Fe_{27}Co_{27}Ni_{27}Si_{10-x}B_9La_x$ ($x = 0, 0.6, 1$) alloys.

The current results thus show that La element substitution for Si has a great influence on the crystallization temperature of the alloy, and it is interesting that La can inhibit the formation of intermetallic compounds. Furthermore, under the same heat treatment conditions, the saturation magnetization of the strip with a certain La content is higher, while the coercivity is relatively lower.

4. Conclusions

A new type of soft magnetic $Fe_{27}Co_{27}Ni_{27}Si_{10-x}B_9La_x$ HEAs were developed in this work. The effects of La on the phase stability, amorphous forming ability and magnetic properties of Fe–Co–Ni–Si–B HE-MGs were investigated. It was found that the soft magnetic properties of $Fe_{27}Co_{27}Ni_{27}Si_{10-x}B_9La_x$ HEAs can be effectively tailored by adjusting their phase structure by annealing treatment. These alloys exhibit a low H_c and a high B_s, in which the values are less than 25 A/m and higher than 1.0 T, respectively. The La content has an important effect on the values of T_{x1} and T_{x2} of the alloys. By increasing the annealing temperature, these alloys precipitated the BCC phase at the first crystallization temperature and transformed into the FCC phase at the phase transition temperature. In additionally, La can inhibit the formation of intermetallic compounds at high temperatures. This work suggests that an optimized annealing temperature is required to obtain good combination of the soft magnetic properties for HEAS.

Author Contributions: Conceptualization, J.L. and J.Z.; methodology and data curation, J.L.; writing-original draft preparation, J.L.; supervision, H.Y. All authors have read and agreed to the published version of the manuscript.

Funding: This work was supported by Guangdong Provincial Natural Science Foundation of China (No. 2021A1515010642).

Acknowledgments: All individuals included in this section have consented to the acknowledgement.

Conflicts of Interest: The authors declare no conflict of interest.

References

1. Cantor, B.; Chang, I.T.H.; Knight, P.; Vincent, A.J.B. Microstructural development in equiatomic multicomponent alloys. *Mater. Sci. Eng.* **2004**, *375–377*, 213–218. [CrossRef]
2. Yeh, J.W.; Chen, S.K.; Lin, S.J.; Gan, J.-Y.; Chin, T.-S.; Shun, T.; Tsau, C.-H.; Chang, S.Y. Nanostructured high-entropy alloys with multiple principal elements: Novel alloy design concepts and outcomes. *Adv. Eng. Mater.* **2004**, *6*, 299–303. [CrossRef]
3. Yeh, J.-W. Recent progress in high-entropy alloys. *Ann. Chim. Sci. Matériaux* **2006**, *31*, 633–648. [CrossRef]
4. *High-Entropy Alloys Fundamentals and Applications*; Springer: Berlin/Heidelberg, Germany, 2016.
5. Yeh, J.-W.; Chang, S.-Y.; Hong, Y.-D.; Chen, S.-K.; Lin, S.-J. Anomalous decrease in X-ray diffraction intensities of Cu–Ni–Al–Co–Cr–Fe–Si alloy systems with multi-principal elements. *Mater. Chem. Phys.* **2007**, *103*, 41–46. [CrossRef]
6. Kukshal, V.; Patnaik, A.; Bhat, I.K. Corrosion and thermal behaviour of AlCr1.5CuFeNi2Tix high-entropy alloys. *Mater. Today: Proc.* **2018**, *5*, 17073–17079. [CrossRef]
7. Shun, T.T.; Hung, C.H.; Lee, C.F. Formation of ordered/disordered nanoparticles in FCC high entropy alloys. *J. Alloy Compd.* **2010**, *493*, 105–109. [CrossRef]
8. Miracle, D.B. High entropy alloys as a bold step forward in alloy development. *Nat. Commun.* **2019**, *10*, 1805. [CrossRef] [PubMed]
9. Zhang, Y.; Zuo, T.T.; Tang, Z.; Gao, M.C.; Dahmen, K.A.; Liaw, P.K.; Lu, Z.P. Microstructures and properties of high-entropy alloys. *Prog. Mater. Sci.* **2014**, *61*, 1–93. [CrossRef]
10. Chang, C.T.; Shen, B.L.; Inoue, A. Co–Fe–B–Si–Nb bulk glassy alloys with superhigh strength and extremely low magnetostriction. *Appl. Phys. Lett.* **2006**, *88*, 011901. [CrossRef]
11. Yang, X.; Zhang, Y. Prediction of high-entropy stabilized solid-solution in multi-component alloys. *Mater. Chem. Phys.* **2012**, *132*, 233–238. [CrossRef]
12. Jiang, Q.; Zhang, G.; Chen, L.; Wu, J.; Zhang, H.; Jiang, J. Glass formability, thermal stability and mechanical properties of La-based bulk metallic glasses. *J. Alloys Compd.* **2006**, *424*, 183–186. [CrossRef]
13. Jiang, Q.K.; Zhang, Q.G.; Yang, L.; Wang, X.; Saksl, K.; Franz, H.; Wunderlich, R.; Fecht, H.; Jiang, J. La-based bulk metallic glasses with critical diameter up to 30mm. *Acta. Mater.* **2007**, *55*, 4409–4418. [CrossRef]
14. Zheng, Z.; Zhao, G.; Xu, L.; Wang, L.; Yan, B. Influence of Ni addition on nanocrystallization kinetics of FeCo-based amorphous alloys. *J. Non-Cryst. Solids* **2016**, *434*, 23–27. [CrossRef]
15. Miracle, D.B.; Senkov, O.N. A critical review of high entropy alloys and related concepts. *Acta. Mater.* **2017**, *122*, 448–511. [CrossRef]
16. Wei, R.; Sun, H.; Chen, C.; Han, Z.; Li, F. Effect of cooling rate on the phase structure and magnetic properties of Fe26.7Co28.5Ni28.5Si4.6B8.7P3 high entropy alloy. *J. Magn. Magn. Mater.* **2017**, *435*, 184–186. [CrossRef]
17. Xu, Y.; Li, Y.; Zhu, Z.; Zhang, W. Formation and properties of Fe25Co25Ni25(P, C, B, Si)25 high-entropy bulk metallic glasses. *J. Non-Cryst. Solids* **2018**, *487*, 60–64. [CrossRef]
18. Li, Y.; Zhang, W.; Qi, T. New soft magnetic Fe25Co25Ni25(P, C, B)25 high entropy bulk metallic glasses with large supercooled liquid region. *J. Alloys Compd.* **2017**, *693*, 25–31. [CrossRef]

19. Qi, T.; Li, Y.; Takeuchi, A.; Xie, G.; Miao, H.; Zhang, W. Soft magnetic Fe25Co25Ni25(B, Si)25 high entropy bulk metallic glasses. *Intermetallics* **2015**, *66*, 8–12. [CrossRef]
20. Chrobak, A.; Nosenko, V.; Haneczok, G.; Boichyshyn, L. Effect of rare earth additions on magnetic properties of Fe82Nb2B14RE2 (RE = Y, Gd, Tb and Dy) amorphous alloys. *Mater. Chem. Phys.* **2011**, *130*, 603–608. [CrossRef]
21. Zheng, G.T.; Jiang, D.G. Magnetic Induction Effect of Rare-Earth La Modified FeSiB Amorphous Ribbon. *Adv. Mater. Res.* **2011**, *415–417*, 566–570.
22. Zuo, T.; Li, R.B.; Ren, X.J.; Zhang, A. Effects of Al and Si addition on the structure and properties of CoFeNi equal atomic ratio alloy. *J. Magn. Magn. Mater.* **2014**, *371*, 60–68. [CrossRef]
23. Zhang, Y.; Zhou, Y.J.; Lin, J.P.; Chen, G.L.; Liaw, P.K. Solid-Solution Phase Formation Rules for Multi-component Alloys. *Adv. Eng. Mater.* **2008**, *10*, 534–538. [CrossRef]
24. Guo, S.; Liu, C.T. Phase stability in high entropy alloys: Formation of solid-solution phase or amorphous phase. *Prog. Nat. Sci. Mater. Int.* **2011**, *21*, 433–446. [CrossRef]
25. Guo, S.; Ng, C.; Lu, J.; Liu, C.T. Effect of valence electron concentration on stability of fcc or bcc phase in high entropy alloys. *J. Appl. Phys.* **2011**, *109*, 103505. [CrossRef]
26. Zhang, T.; Li, R.; Pang, S. Effect of similar elements on improving glass-forming ability of La–Ce-based alloys. *J. Alloys Compd.* **2009**, *483*, 60–63. [CrossRef]
27. Turnbull, D. Under what conditions can a glass be formed? *Contemp. Phys.* **1969**, *10*, 473–488. [CrossRef]
28. Schroers, J. The Superplastic Forming of Bulk Metallic Glasses. *J. Miner. Met. Mater. Soc.* **2005**, *57*, 35–39. [CrossRef]
29. Zhang, Z.; Song, K.; Li, R.; Xue, Q.; Wu, S.; Yan, D.; Li, X.; Song, B.; Sarac, B.; Kim, J.T.; et al. Polymorphic Transformation and Magnetic Properties of Rapidly Solidified Fe26.7Co26.7Ni26.7Si8.9B11.0 High-Entropy Alloys. *Materials* **2019**, *12*, 590. [CrossRef] [PubMed]
30. Byshkin, M.; Hou, M. Phase transformations and segregation in Fe–Ni alloys and nanoalloys. *J. Mater. Sci.* **2012**, *47*, 5784–5793. [CrossRef]

Article

Surface Investigation of Ni$_{81}$Fe$_{19}$ Thin Film: Using ARXPS for Thickness Estimation of Oxidation Layers

Zongsheng He [1], Ziyu Li [1], Xiaona Jiang [1], Chuanjian Wu [1], Yu Liu [1], Xinglian Song [2], Zhong Yu [1], Yifan Wang [1], Zhongwen Lan [1] and Ke Sun [1,*]

[1] School of Materials and Energy, University of Electronic Science and Technology of China, Chengdu 610054, China; hezongsheng112@163.com (Z.H.); ziyu.li@uestc.edu.cn (Z.L.); xnjiang@uestc.edu.cn (X.J.); wcjuestc2005@uestc.edu.cn (C.W.); liuyui15005510@163.com (Y.L.); yuzhong@uestc.edu.cn (Z.Y.); complax@163.com (Y.W.); zwlan@uestc.edu.cn (Z.L.)
[2] Shandong Chunguang Magnetoelectricity Technology Co., Ltd., Linyi 276017, China; kongxianjuan@ktong.com
* Correspondence: ksun@uestc.edu.cn; Tel.: +86-028-8320-1637

Citation: He, Z.; Li, Z.; Jiang, X.; Wu, C.; Liu, Y.; Song, X.; Yu, Z.; Wang, Y.; Lan, Z.; Sun, K. Surface Investigation of Ni$_{81}$Fe$_{19}$ Thin Film: Using ARXPS for Thickness Estimation of Oxidation Layers. *Metals* **2021**, *11*, 2061. https://doi.org/10.3390/met11122061

Academic Editor: Volodymyr A. Chernenko

Received: 29 November 2021
Accepted: 16 December 2021
Published: 20 December 2021

Publisher's Note: MDPI stays neutral with regard to jurisdictional claims in published maps and institutional affiliations.

Copyright: © 2021 by the authors. Licensee MDPI, Basel, Switzerland. This article is an open access article distributed under the terms and conditions of the Creative Commons Attribution (CC BY) license (https://creativecommons.org/licenses/by/4.0/).

Abstract: This work demonstrates the dependence between magnetic properties and the thickness of NiFe thin films. More importantly, a quantitative study of the surface composition of NiFe thin film exposed to atmospheric conditions has been carried out employing angle-resolved X-ray photoelectron spectroscopy (ARXPS). In this study, we fabricated Ni$_{81}$Fe$_{19}$ (NiFe) thin films on Si (100) substrate using electron beam evaporation and investigated their surface morphologies, magnetic properties, and the thickness of the surface oxide layer. The coexistence of metallic and oxidized species on the surface are suggested by the depth profile of ARXPS spectra. The thickness of the oxidized species, including NiO, Ni(OH)$_2$, Fe$_2$O$_3$, and Fe$_3$O$_4$, are also estimated based on the ARXPS results. This work provides an effective approach to clarify the surface composition, as well as the thickness of the oxide layer of the thin films.

Keywords: permalloy; magnetic thin films; ARXPS; magnetic property; oxidation layer

1. Introduction

Microwave magnetic devices, such as circulators, filters, and phase shifters are indispensable components in satellite and mobile communications systems [1–3]. Recently, the rapid development in the electronic communication industry has proposed higher requirements for microwave devices including higher operating frequency, lower loss, and higher integration level. As magnetoelectronic devices are the core components of microwave devices [4–6], there is an urgent demand to improve their magnetic properties at the high-frequency range. However, most microwave/radio-frequency magnetic components are discrete devices based on bulk materials, and directly reducing the size of high integration often leads to degradation in performance. The progress of chip-type technology is still behind for the perfect integration of magnetic components with existing semiconductor devices. Given the above context, ferromagnetic/anti-ferromagnetic (FM/AF) thin films with a multilayer structure are considered as ideal candidates. The higher saturation magnetization (M_s), permeability (μ), and self-biased ferromagnetic resonance (FMR) frequency of the magnetic thin film make them suitable for high-frequency applications as nanostructured magnetic media and magnetic sensors [7–13]. Meanwhile, the multilayer structure has been widely used in giant magnetoresistance spin valves based on the exchange bias phenomenon.

The operating frequency, FMR linewidth and the resonance field of periodically arranged FM/AF multilayer structure can easily be affected by the change in thickness of each single layer, as well as the surface morphology. Especially, when the thickness of individual layers falls below the critical thickness, the states and properties of surface and interface dominate the magnetic properties. In the multilayer structure of thin films, the thickness of each single layer can reach down to the level of several nanometers. However, such high accuracy is often susceptible to environmental oxidation, which leads to the change in the effective thickness, as well as the change in composition.

To study the composition of surface and interface as well as the thickness distribution in multilayer structures, atomic force microscopy (AFM) and angle resolved X-ray photoelectron spectroscopy (ARXPS) are commonly applied. Among the previous studies, Yu G.H. et al. [14] studied the relationship between the oxidation states of Ni in NiO_x and the exchange bias field H_{ex} of the NiFe film by ARXPS, but the critical thickness of the pinning of NiFe film was not studied. S.S. Sakhonenkov et al. [15] presented an in-depth study of Mo/Si multilayer systems using ARXPS. They reported that a $MoSi_2$ interlayer with a thickness of 0.19 ± 0.05 nm was identified on the Si-on-Mo interface. J. Zemek et al. [16] investigated the surface and in-depth distribution of sp^2 and sp^3 coordinated carbon atoms in modified diamond-like carbon films. A. Sanchez-Martinez et al. [17] discovered that small amounts of oxidized gallium and metallic arsenic are located at the HfO_2/InGaAs interface. Meanwhile, they studied the structure of TiN/HfO_2 nanofilms grown on InGaAs substrates by ARXPS. The above evidences have shown ARXPS as a useful tool in characterizing the surface/interface of the thin films. In this study, AFM and ARXPS are used to characterize the surface oxidation layers of the NiFe thin film, which is known for its high permeability while having a relatively high M_s. More importantly, the thickness of each oxidation layer is estimated based on the ARXPS results. The static magnetic properties of the NiFe thin films were also investigated.

2. Experimental Details

In this experiment, $Ni_{81}Fe_{19}$ thin films were deposited on 5 mm × 5 mm Si (100) substrates by electron beam evaporation (EB-500) in vacuum ($\leq 5.0 \times 10^{-5}$ Pa) at 25 °C. The thicknesses of the $Ni_{81}Fe_{19}$ films are 90, 100, 110, and 120 nm. The growth rate of the film is 0.03 nm/s. The Si (100) substrates were preliminarily cleaned in a sequential bath of acetone, alcohol, and deionized water and dried with ionized dry N_2 flux. Once prepared, the samples were stored at room temperature to be oxidized under ambient environment.

The surface morphologies of the NiFe thin films were investigated using an atomic force microscope (AFM, Bruker MultiMode8) with a scanned area of 2 μm × 2 μm. Static magnetic properties of NiFe films were measured using a vibrating sample magnetometer (VSM, Lake Shore 8604) at room temperature. The samples were analyzed using ARXPS (ULVAC-PHI5000 Veraprobe III) with an Al Kα emission source (1486.6 eV). The XPS measurements were conducted at a base pressure of 8×10^{-10} Pa. The pass energy of the spectrometer was set to 69 eV. The energy was calibrated by setting the adventitious C 1s binding energy to 284.8 eV. The chemical depth profile was acquired at six angles (α) of 20°, 35°, 45°, 60°, 75°, and 90°.

3. Results and Discussion

3.1. AFM Analysis and Surface Roughness

Figure 1 illustrates the AFM surface morphology and roughness of the NiFe thin films with a NiFe thicknesses of 100 nm. The surface roughness of the films with different NiFe thicknesses from 90 to 120 nm was 1.31 nm, 1.27 nm, 1.45nm, and 1.67 nm, respectively. The minimum roughness of 1.27 nm is seen for the sample with a thickness of 100 nm, whose average grain size also appears to be the smallest among the samples (see Figure 3a). The experimental results are in good correspondence with a previous report [18], which shows the positive correlation between the average grain size and surface roughness.

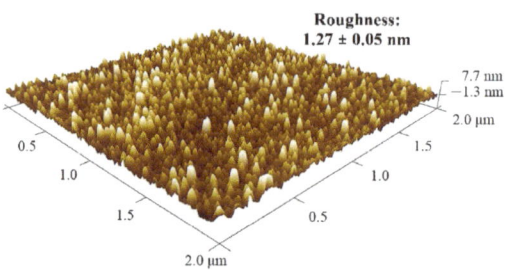

Figure 1. AFM images of NiFe films with a thickness of 100 nm.

3.2. Magnetic Properties of NiFe Films

Figure 2a,b shows the in-plane and out-of-plane magnetic hysteresis loops of NiFe thin films with the thicknesses from 90 to 120 nm. The samples with thickness of 90–110 nm show narrow M-H loops, while the 120-nm-thick sample shows a much wider curve, suggesting a higher coercivity (H_c) [19]. Figure 3 shows the static magnetic properties of the NiFe thin films, including the saturation magnetization ($4\pi M_s$) and H_c. According to Figure 3a, as the thickness of NiFe thin film increases, the $4\pi M_s$ is increased. Concomitantly, in-plane ($H_{c//}$) and out-of-plane ($H_{c\perp}$) coercivity are displayed in Figure 3b. Both $H_{c//}$ and $H_{c\perp}$ exhibit a similar trend, which shows a slight reduction upon the increase of thickness from 90 to 110 nm, followed by a dramatic increase as the thickness is further increased to 120 nm. As the H_c is dependent on the quality and defects of the film [20], the dramatic increase of H_c observed in Figure 3b can be attributed to the high roughness of the 120 nm thick sample.

3.3. Oxidation Thickness of NiFe Films

Figure 4 shows the ARXPS spectra of Ni in the NiFe thin film with a thickness of 100 nm from various take-off angles. For clear identification of each peak, as well as to obtain the peak intensity, fitted results for each spectrum are obtained (examples of $\alpha = 20°$ and $90°$ are shown in Figure 5). The relationship between the detection depth (d), photoelectron take-off grazing angle, and the mean free path of inelastic scattering (λ), is as follows [21].

$$d = 3\lambda \sin \alpha \tag{1}$$

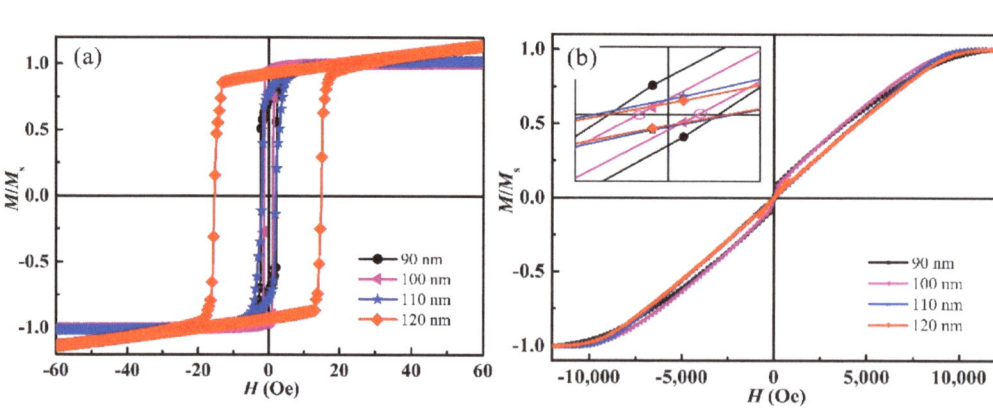

Figure 2. The magnetic hysteresis (M-H) loops of NiFe thin films with thickness from 90 to 120 nm: (**a**) the in-plane magnetic hysteresis loops, (**b**) the out-of-plane magnetic hysteresis loop.

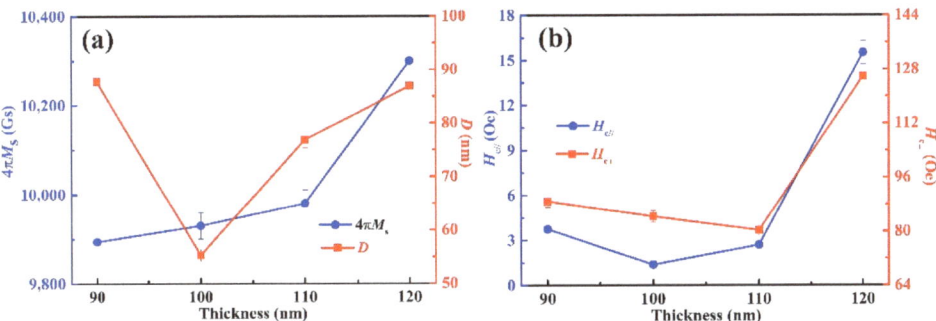

Figure 3. The static magnetic properties of the Ni$_{81}$Fe$_{19}$ thin films with thicknesses from 90 to 120 nm: (**a**) the saturation magnetization (4πM_s) and average grain size (*D*), (**b**) the in-plane ($H_{c//}$) and out-of-plane ($H_{c\perp}$) coercivity.

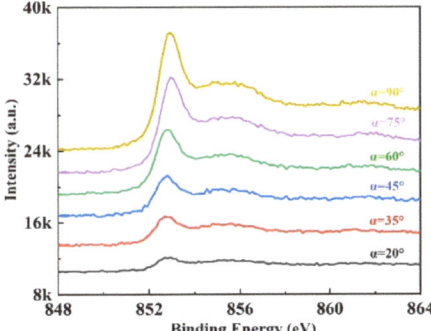

Figure 4. Spectra of Ni 2p photoemission in the NiFe thin film with a thickness of 100 nm at different take-off angles (α).

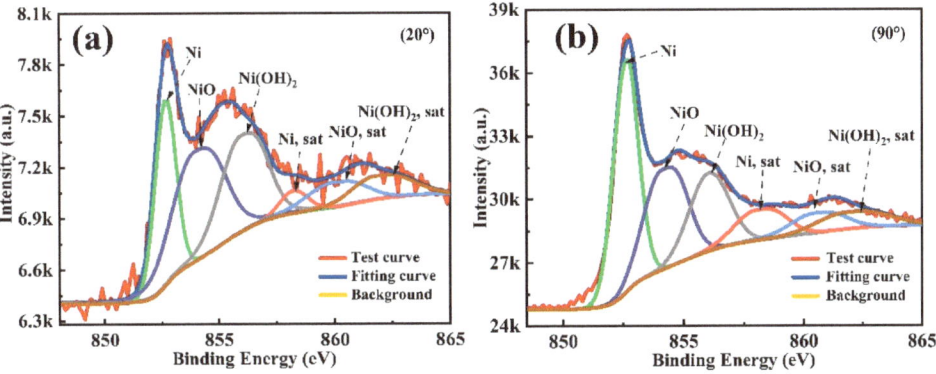

Figure 5. ARXPS spectra of Ni 2p photoemission and the fitting results for NiFe thin film with a thickness of 100 nm at different take-off angles: (**a**) 20° and (**b**) 90°.

As can be seen in Figure 5, there are several noticeable peaks representing the signals from Ni 2p3/2 (852.3 eV), NiO 2p3/2 (853.3 eV), Ni(OH)$_2$ 2p3/2 (856.6 eV), as well as their satellite peaks [22–24]. The coexistence of metallic and oxide components is further confirmed by the spectra from O 1s photoemission, which is shown in Figure 6a. The thickness of the oxide film (d_o) can be calculated according to the ARXPS results through the following equations [25]:

$$\ln\left(1+\frac{R}{R_\infty}\right) = \frac{d_o}{\lambda} \cdot \frac{1}{\sin\alpha} \tag{2}$$

where α is different take-off angles: 20°, 35°, 45°, 60°, 75°, and 90°, and R is the ratio of photoelectron peak intensity from the oxides (I_o) to that from nickel (I_s):

$$R = \frac{I_o}{I_s} \tag{3}$$

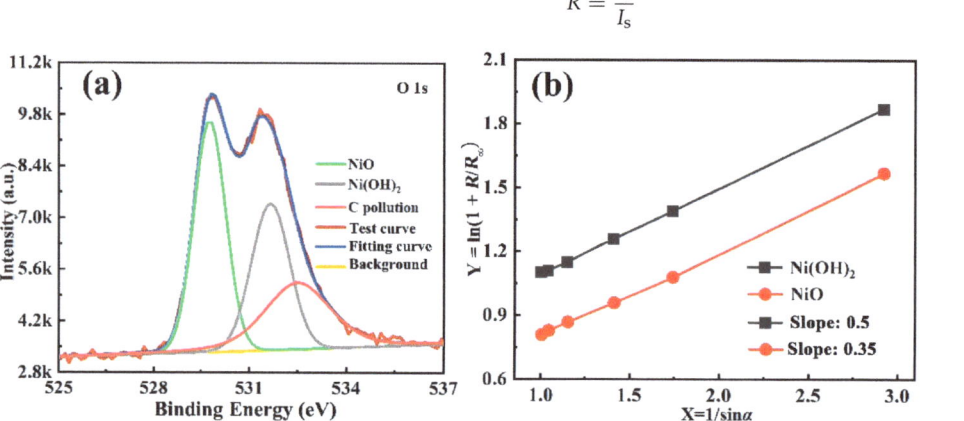

Figure 6. (a) ARXPS spectra of O 1s photoemission and the fitting results, (b) the results of $\ln(1 + R/R_\infty)$ against $1/\sin\alpha$ for the nickel oxides in NiFe thin film.

In Equation (2), R_∞ can be calculated as follows:

$$R_\infty = \frac{M_o}{M_s} \cdot \frac{\rho_s}{\rho_o} \cdot \frac{\lambda_s}{\lambda_o} \tag{4}$$

where the values of M are the molecular weight of nickel oxides (M_o) and nickel (M_s), values of ρ are the density of nickel oxides (ρ_o) and nickel (ρ_s), and values of λ are the mean free path of inelastic scattering for nickel oxides (λ_o) and nickel (λ_s). Because the photoelectron kinetic energy of nickel and its oxides are almost the same, thus λ_o and λ_s are approximately equal [21]. Table 1 shows the values of density, molecular weight for Ni, NiO, and Ni(OH)$_2$, as well as the calculated R_∞ for both nickel oxides.

Table 1. The values of density, molecular weight, and kinetic energy for Ni, NiO, and Ni(OH)$_2$, as well as R_∞ for both nickel oxides.

Material	ρ (g/cm^3)	M (g/mol)	R_∞	Kinetic Energy (eV)
Ni	8.9	59	-	853.3
NiO	6.84	75	0.6	853.6
Ni(OH)$_2$	4.15	93	0.3	855.0

Table 2 shows the relative peak intensities of nickel (I_s) and nickel oxides (I_o) at different take-off angles based on the fitting results shown in Figure 5, as well as the calculated ratios between them (R). As α is increased from 20 to 90°, I_s(Ni) is seen to increase from ~20 to ~43%, while both I_o(NiO) and I_o(Ni(OH)$_2$) are reduced from ~46 to ~32% and from ~33 to ~26%, respectively. As the oxidation reactions take place at the top surface of the metal substrate, with a limited traveling distance of photoemission signal, it is natural that the detection depth becomes deeper as the take-off angle becomes closer to

90°. Therefore, one can observe the obvious enhancement in I_s(Ni), and slight reduction in I_o(NiO) and I_o(Ni(OH)$_2$).

Table 2. Relative peak intensities (I_s and I_o) and the I_o/I_s ratios for both nickel oxides in the NiFe (100 nm) thin film.

α (°)	I_s(Ni) (%)	I_o(NiO) (%)	I_o(Ni(OH)$_2$) (%)	R(NiO) = I_o(NiO)/I_s	R(Ni(OH)$_2$) = I_o/I_s
20	20.38	46.18	33.44	2.27	1.64
35	32.50	37.95	29.55	1.17	0.91
45	36.73	35.37	27.90	0.96	0.76
60	40.32	33.27	26.41	0.83	0.66
75	42.05	32.29	25.66	0.77	0.61
90	42.57	32.00	25.44	0.75	0.60

According to Equation (2), to calculate the thickness of the oxidation layer, one can plot the results of $\ln(1 + R/R_\infty)$ against $1/\sin\alpha$, the slope of which indicates the thickness divided by the mean free path of inelastic scattering (d_o/λ). Figure 6b shows the results of $\ln(1 + R/R_\infty)$ against $1/\sin\alpha$ for Ni(OH)$_2$ and NiO. Given that $\lambda = 6$ Å [26], the thickness of NiO in NiFe thin film is determined to be 0.2 nm, the thickness of Ni(OH)$_2$ is determined to be 0.3 nm.

As with nickel oxides, the composition and thickness of iron oxides can also be determined. Figure 7 shows the Fe 2p photoemission spectra of partially oxidized NiFe thin film. The signals from each take-off angle can be fitted with Fe 2p3/2 (706.8 eV), Fe$_2$O$_3$ 2p3/2 (710.7 eV), and Fe$_3$O$_4$ 2p3/2 (709.3 eV) [27–29], the results of which are shown in Figure 8. Figure 9a shows the fitted O 1s spectra to further confirm the existence of iron oxides. Table 3 shows the relevant parameters including I_s(Fe), I_o(Fe$_2$O$_3$), I_o(Fe$_3$O$_4$), and the intensity ratios between iron and its oxides. For the same reason mentioned above, I_s(Fe) is increased with increasing α, as higher take-off angle results in deeper detection depth.

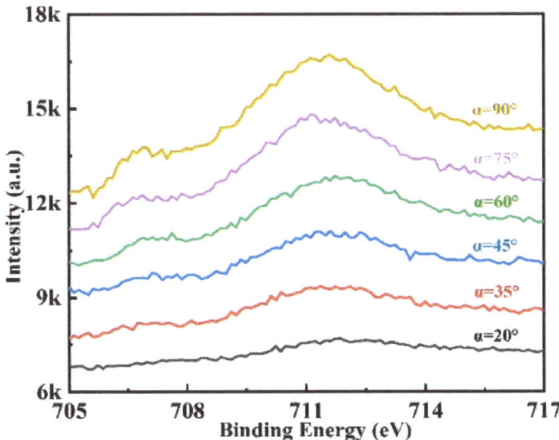

Figure 7. Spectra at different take-off angle of Fe 2p in NiFe(100 nm) thin film.

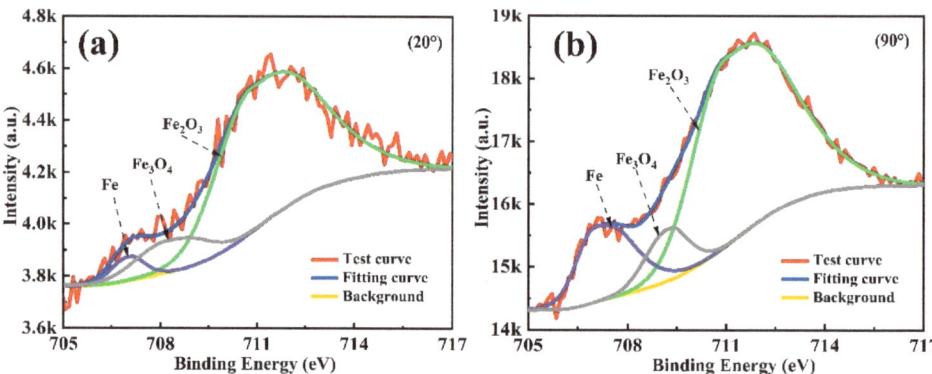

Figure 8. ARXPS spectra of Fe 2p photoemission and the peak fitting results for NiFe thin film with a thickness of 100 nm at different take-off angles of: (**a**) 20° and (**b**) 90°.

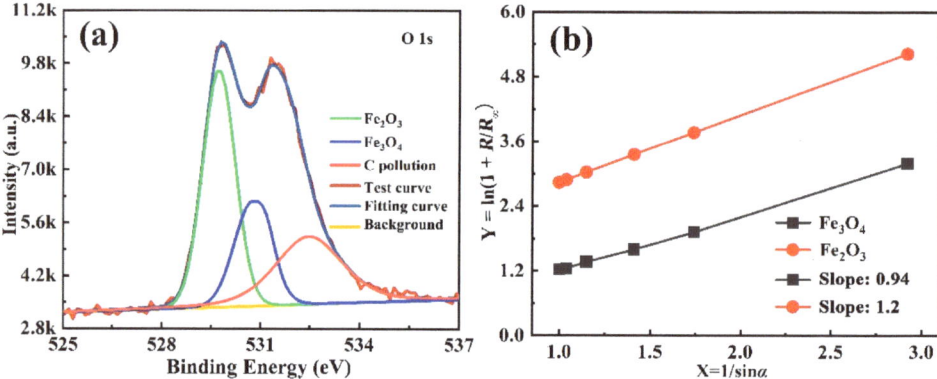

Figure 9. (**a**) ARXPS spectra of O 1s photoemission and the fitting results, (**b**) the results of $\ln(1 + R/R_\infty)$ against $1/\sin\alpha$ for the iron oxides in NiFe thin film.

Table 3. Relative peak intensities (I_s and I_o) and the I_o/I_s ratios for both iron oxides in the NiFe (100 nm) thin film.

α (°)	I_s(Fe) (%)	I_o(Fe$_3$O$_4$) (%)	I_o(Fe$_2$O$_3$) (%)	R(Fe$_3$O$_4$) = I_o(Fe$_3$O$_4$)/I_s	R(Fe$_2$O$_3$) = I_o(Fe$_2$O$_3$)/I_s
20	2.11	7.85	90.04	3.72	42.67
35	8.59	8.01	83.41	0.93	9.71
45	12.49	7.92	79.58	0.63	6.37
60	16.63	7.75	75.61	0.47	4.55
75	18.91	7.63	73.46	0.40	3.88
90	19.63	7.59	72.78	0.39	3.71

Table 4 shows the values of density, molecular weight for Fe, Fe$_3$O$_4$, and Fe$_2$O$_3$, as well as the calculated R_∞ for both iron oxides. The values in Tables 3 and 4 are integrated into Equations (2)–(4), and the results of $\ln(1 + R/R_\infty)$ against $1/\sin\alpha$ for Fe 2p photoemission in the NiFe thin film are shown in Figure 9b. As λ = 7.5 Å [26], the thickness of Fe$_2$O$_3$ layer determined is 0.9 nm and that of Fe$_3$O$_4$ is 0.7 nm. It means that oxidation in the natural environment occurs only in the extreme depth of the surface. To summarize, a schematic illustration is shown in Figure 10 to better demonstrate the multilayer structure of the oxidized NiFe.

Table 4. The values of density, molecular weight, and kinetic energy for Fe, Fe$_2$O$_3$, and Fe$_3$O$_4$, as well as R$_\infty$ for both iron oxides.

Material	ρ (g/cm^3)	M (g/mol)	R_∞	Kinetic Energy (eV)
Fe	7.86	56	-	710.7
Fe$_2$O$_3$	5.24	160	0.23	710.4
Fe$_3$O$_4$	5.18	232	0.16	710.8

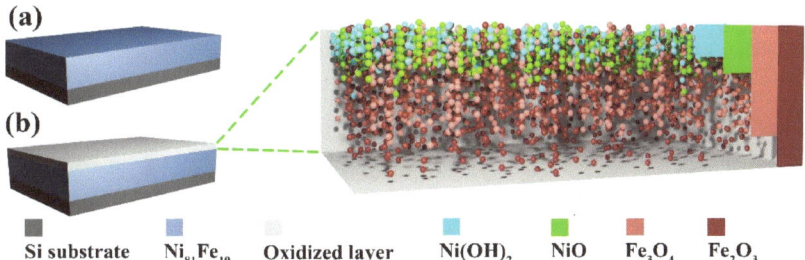

Figure 10. The schematic diagram of NiFe thin film: (**a**) unoxidized, (**b**) oxidized.

4. Conclusions

In summary, NiFe thin films with thickness from 90 to 120 nm were fabricated by electron beam evaporation. The sample with a thickness of 100 nm shows the lowest surface roughness and the smallest grain size. The saturation magnetization of the films is increased from 9891 to 10,300 Gs as the thickness of film increases. The optimum magnetic properties ($4\pi M_s$ = 9930 Gs, H_c = 1.37 Oe) are obtained when the thickness of film is 100 nm. In addition, the coexistence of NiFe and its oxides on the top surface is observed along the probing depth of ARXPS. By fitting the ARXPS spectra and obtaining the relative intensity of each peak, the thickness of NiO, Ni(OH)$_2$, Fe$_2$O$_3$, and Fe$_3$O$_4$, which are 0.3 nm, 0.2 nm, 0.9 nm, and 0.7 nm, respectively, are determined. This work demonstrates the beneficial application of ARXPS in characterizing the oxide layers including their compositions and thicknesses as a quick and damage-free approach, especially for future investigations with even lower film thickness.

Author Contributions: Conceptualization, K.S.; funding acquisition, K.S.; investigation, Z.H., X.S. and X.J.; methodology, Z.H.; software, Y.L. and Y.W.; supervision, C.W., Z.Y., Z.L. (Zhongwen Lan) and K.S.; writing—original draft, Z.H.; writing—review and editing, Z.H. and Z.L. (Ziyu Li). All authors have read and agreed to the published version of the manuscript.

Funding: This present work was financially supported by the National Natural Science Foundation of China under grant No. 52172267 and 51772046.

Institutional Review Board Statement: Not applicable.

Informed Consent Statement: Not applicable.

Data Availability Statement: Not applicable.

Conflicts of Interest: The authors declare that they have no known competing financial interests or personal relationships that could have appeared to influence the work reported in this paper.

References

1. Su, Z.; Bennett, S.; Hu, B.; Chen, Y.; Harris, V.G. Magnetic and microwave properties of U-type hexaferrite films with high remanence and low ferromagnetic resonance linewidth. *J. Appl. Phys.* **2014**, *115*, 17A504. [CrossRef]
2. Belyaev, B.A.; Lemberg, K.V.; Serzhantov, A.M.; Leksikov, A.A.; Bal'va, Y.F. Magnetically tunable resonant phase shifters for UHF band. *IEEE Trans. Magn.* **2015**, *51*, 1–5. [CrossRef]
3. Sharma, V.; Khivintsev, Y.; Harward, I.; Kuanr, B.K.; Celinski, Z. Fabrication and characterization of microwave phase shifter in microstrip geometry with Fe film as the frequency tuning element. *J. Magn. Magn. Mater.* **2019**, *489*, 165412. [CrossRef]

4. Kuanr, B.K.; Veerakumar, V.; Camley, R.E.; Celinski, Z. Permalloy (NiFe) nanometer square-antidot arrays: Dynamic modes and use as a monolithic microwave band-pass filter. *J. Magn. Magn. Mater.* **2019**, *484*, 272–278. [CrossRef]
5. Lu, Q.; Yu, B.; Hu, Z.; He, Y.; Hu, T.; Zhao, Y.; Wang, Z.; Zhou, Z.; Cui, W.; Liu, M. Surface roughness evolution induced low secondary electron yield in carbon coated Ag/Al substrates for space microwave devices. *Appl. Surf. Sci.* **2020**, *501*, 144236. [CrossRef]
6. Paiva, D.V.M.; Silva, M.A.S.; Sombra, A.S.B.; Fechine, P.B.A. Properties of the Sr_3MoO_6 electroceramic for RF/microwave devices. *J. Alloys Compd.* **2018**, *748*, 766–773. [CrossRef]
7. Chen, Y.; Fan, X.; Zhou, Y.; Xie, Y.; Wu, J.; Wang, T.; Chui, S.T.; Xiao, J.Q. Designing and tuning magnetic resonance with exchange interaction. *Adv. Mater.* **2015**, *27*, 1351–1355. [CrossRef]
8. Belmeguenai, M.; Martin, T.; Woltersdorf, G.; Bayreuther, G.; Baltz, V. Microwave spectroscopy with vector network analyzer for interlayer exchange-coupled symmetrical and asymmetrical NiFe/Ru/NiFe. *J. Phys. Condens. Matter.* **2008**, *20*, 345206. [CrossRef]
9. Li, S.; Li, Q.; Xu, J.; Yan, S.; Miao, G.X.; Kang, S.; Dai, Y. Tunable optical mode ferromagnetic resonance in FeCoB/Ru/FeCoB synthetic antiferromagnetic trilayers under uniaxial magnetic anisotropy. *Adv. Funct. Mater.* **2016**, *26*, 3738–3744. [CrossRef]
10. Xu, F.; Liao, Z.; Huang, Q.; Phuoc, N.N.; Ong, C.K.; Li, S. Influence of thickness on magnetic properties and microwave characteristics of NiFe/IrMn/NiFe Trilayers. *IEEE Trans. Magn.* **2011**, *47*, 3486–3489. [CrossRef]
11. Naik, R.; Kota, C.; Payson, J.S.; Dunifer, G.L. Ferromagnetic-resonance studies of epitaxial Ni, Co, and Fe films grown on Cu(100)/Si(100). *Phys. Rev. B* **1993**, *48*, 1008–1013. [CrossRef]
12. Chappert, C.; Le Dang, K.; Beauvillain, P.; Hurdequint, H.; Renard, D. Ferromagnetic resonance studies of very thin cobalt films on a gold substrate. *Phys. Rev. B* **1986**, *34*, 3192–3197. [CrossRef]
13. Biondo, A.; Nascimento, V.P.; Lassri, H.; Passamani, E.C.; Morales, M.A.; Mello, A. Structural and magnetic properties of $Ni_{81}Fe_{19}$/Zr multilayers. *J. Magn. Magn. Mater.* **2004**, *277*, 144–152. [CrossRef]
14. Yu, G.; Zhao, H.; Teng, J.; Chai, C.; Zhu, F.; Xia, Y.; Chai, X. XPS Studies of magnetic multilayers. *J. Univ. Sci. Technol. B* **2001**, *8*, 210–213.
15. Sakhonenkov, S.S.; Filatova, E.O.; Gaisin, A.U.; Kasatikov, S.A.; Konashuk, A.S. Angle resolved photoelectron spectroscopy as applied to X-ray mirrors: An in-depth study of Mo/Si multilayer systems. *Phys. Chem. Chem. Phys.* **2019**, *21*, 25002–250109. [CrossRef]
16. Zemek, J.; Houdkova, J.; Jiricek, P.; Jelinek, M. Surface and in-depth distribution of sp^2 and sp^3 coordinated carbon atoms in diamond-like carbon films modified by argon ion beam bombardment during growth. *Carbon* **2018**, *134*, 71–79. [CrossRef]
17. Sanchez-Martinez, A.; Ceballos-Sanchez, O.; Vazquez-Lepe, M.O.; Duong, T.; Arroyave, R. Diffusion of In and Ga in TiN/HfO_2/InGaAs nanofilms. *J. Appl. Phys.* **2013**, *114*, 143504. [CrossRef]
18. Dong, B.Z.; Fang, G.J.; Wang, J.F.; Guan, W.J.; Zhao, X.Z. Effect of thickness on structural, electrical, and optical properties of ZnO: Al films deposited by pulsed laser deposition. *J. Appl. Phys.* **2007**, *101*, 033713. [CrossRef]
19. Akhter, M.A.; Mapps, D.J.; Ma Tan, Y.Q.; Petford-Long, A.; Doole, R. Thickness and grain-size dependence of the coercivity in permalloy thin films. *J. Appl. Phys.* **1997**, *81*, 4122–4124. [CrossRef]
20. Berzins, A.; Smits, J.; Petruhins, A. Characterization of Microscopic ferromagnetic defects in thin films using magnetic microscope based on nitrogen-vacancy centres. *Mater. Chem. Phys.* **2021**, *267*, 124617. [CrossRef]
21. John, F.W.; John, W. *An Introduction to Surface Analysis by XPS and AES*, 1st ed.; East China University of Technology Publisher: Shanghai, China, 2008.
22. Nagarkar, P.V.; Kulkarni, S.K.; Umbach, E. Surface oxidation investigation of $Ni_{36}Fe_{32}Cr_{14}P_{12}B_6$ glass using angle resolved XPS. *Appl. Surf. Sci.* **1987**, *29*, 194–222. [CrossRef]
23. Biesinger, M.C.; Payne, B.P.; Grosvenor, A.P.; Lau, L.W.M.; Gerson, A.R.; Smart, R.S.C. Resolving surface chemical states in XPS analysis of first row transition metals, oxides and hydroxides: Cr, Mn, Fe, Co and Ni. *Appl. Surf. Sci.* **2011**, *257*, 2717–2730. [CrossRef]
24. Dupin, J.C.; Gonbeau, D.; Vinatier, P.; Levasseur, A. Systematic XPS studies of metal oxides, hydroxides and peroxides. *Phys. Chem. Chem. Phys.* **2000**, *2*, 1319–1324. [CrossRef]
25. Wang, J.Q. *Introduction to Electron Spectroscopy (XPS/XAES/UPS)*; National Defense Industry Publisher: Beijing, China, 1992.
26. Cao, Q.X.; Lei, T.M.; Huang, Y.X.; Li, G.F. *Fundamentals of Solid Physics*; Xidian University Publisher: Xi'an, China, 2008.
27. Yamashita, T.; Hayes, P. Analysis of XPS spectra of Fe^{2+} and Fe^{3+} ions in oxide materials. *Appl. Surf. Sci.* **2008**, *254*, 2441–2449. [CrossRef]
28. Grosvenor, A.P.; Kobe, B.A.; Biesinger, M.C.; McIntyre, N.S. Investigation of multiplet splitting of Fe 2p XPS spectra and bonding in iron compounds. *Surf. Interface Anal.* **2004**, *36*, 1564–1574. [CrossRef]
29. Carver, J.C.; Schweitzer, G.K.; Carlson, T.A. Use of X-ray photoelectron spectroscopy to study bonding in Cr, Mn, Fe, and Co compounds. *J. Chem. Phys.* **1972**, *57*, 973–982. [CrossRef]

Article

One-Step Sintering Process for the Production of Magnetocaloric La(Fe,Si)$_{13}$-Based Composites

Xi-Chun Zhong [1,*], Xu-Tao Dong [1], Jiao-Hong Huang [2], Cui-Lan Liu [2], Hu Zhang [3], You-Lin Huang [4], Hong-Ya Yu [1] and Raju V. Ramanujan [5,6]

[1] School of Materials Science and Engineering, South China University of Technology, Guangzhou 510640, China; xutaodonga@gmail.com (X.-T.D.); yuhongya@scut.edu.cn (H.-Y.Y.)
[2] Baotou Research Institute of Rare Earths, Baotou 014030, China; jiaohongh@163.com (J.-H.H.); liucuilan618@126.com (C.-L.L.)
[3] School of Materials Science and Engineering, University of Science and Technology Beijing, Beijing 100083, China; zhanghu@ustb.edu.cn
[4] School of Materials Science and Engineering, Nanchang Hangkong University, Nanchang 330063, China; hyl1019_lin@163.com
[5] School of Materials Science and Engineering, Nanyang Technological University, Singapore 639798, Singapore; ramanujan@ntu.edu.sg
[6] Singapore-HUJ Alliance for Research and Enterprise (SHARE), Nanomaterials for Energy and Energy-Water Nexus (NEW), Campus for Research Excellence and Technological Enterprise (CREATE), Singapore 138602, Singapore
* Correspondence: xczhong@scut.edu.cn

Citation: Zhong, X.-C.; Dong, X.-T.; Huang, J.-H.; Liu, C.-L.; Zhang, H.; Huang, Y.-L.; Yu, H.-Y.; Ramanujan, R.V. One-Step Sintering Process for the Production of Magnetocaloric La(Fe,Si)$_{13}$-Based Composites. *Metals* 2022, *12*, 112. https://doi.org/10.3390/met12010112

Academic Editors: Volodymyr A. Chernenko and Daniel Fruchart

Received: 9 November 2021
Accepted: 4 January 2022
Published: 6 January 2022

Publisher's Note: MDPI stays neutral with regard to jurisdictional claims in published maps and institutional affiliations.

Copyright: © 2022 by the authors. Licensee MDPI, Basel, Switzerland. This article is an open access article distributed under the terms and conditions of the Creative Commons Attribution (CC BY) license (https://creativecommons.org/licenses/by/4.0/).

Abstract: A one-step sintering process was developed to produce magnetocaloric La(Fe,Si)$_{13}$/Ce-Co composites. The effects of Ce$_2$Co$_7$ content and sintering time on the relevant phase transformations were determined. Following sintering at 1373 K/30 MPa for 1–6 h, the NaZn$_{13}$-type (La,Ce)(Fe,Co,Si)$_{13}$ phase formed, the mass fraction of α-Fe phase reduced and the CeFe$_7$-type (La,Ce)(Fe,Co,Si)$_7$ phase appeared. The mass fraction of the (La,Ce)(Fe,Co,Si)$_7$ phase increased, and the α-Fe phase content decreased with increasing Ce$_2$Co$_7$ content. However, the mass fraction of the (La,Ce)(Fe,Co,Si)$_7$ phase reduced with increasing sintering time. The EDS results showed a difference in concentration between Co and Ce at the interphase boundary between the 1:13 phase and the 1:7 phase, indicating that the diffusion mode of Ce is reaction diffusion, while that of Co is the usual vacancy mechanism. Interestingly, almost 100% single phase (La,Ce)(Fe,Co,Si)$_{13}$ was obtained by appropriate Ce$_2$Co$_7$ addition. After 6 h sintering at 1373 K, the Ce and Co content in the (La,Ce)(Fe,Co,Si)$_{13}$ phase increased for larger Ce$_2$Co$_7$ content. Therefore, the Curie temperature increased from 212 K (binder-free sample) to 331 K (15 wt.% Ce$_2$Co$_7$ sample). The maximum magnetic entropy change ($-\Delta S_M$)^{max} decreased from 8.8 (binder-free sample) to 6.0 J/kg·K (15 wt.% Ce$_2$Co$_7$ sample) under 5 T field. High values of compressive strength (σ_{bc})^{max} of up to 450 MPa and high thermal conductivity (λ) of up to 7.5 W/m·K were obtained. A feasible route to produce high quality La(Fe,Si)$_{13}$ based magnetocaloric composites with large MCE, good mechanical properties, attractive thermal conductivity and tunable T_C by a one-step sintering process has been demonstrated.

Keywords: La(Fe,Si)$_{13}$ based composites; sintering; grain boundary diffusion; magnetocaloric effect

1. Introduction

Climate change is of high global significance, resulting in high interest in more efficient cooling systems. Therefore, there is considerable research interest in the emerging attractive topic of magnetic refrigeration (MR) [1,2]. Advances in magnetocaloric materials (MCMs) are urgently needed to realize the promise of MR. MCMs which exhibit a giant magnetocaloric effect (MCE) include Gd$_5$(Si$_x$Ge$_{1-x}$)$_4$ [3,4], MnFe(P,As) [5], Mn(Fe,Co)Ge [6], LaFe$_{13-x}$Si$_x$ [7,8], and Heusler alloys [9], which undergo a first-order magnetic transition (FOMT) near their Curie temperatures (T_Cs). Recently, other materials, such as RE_2ZnMnO_6 (RE = Gd, Dy, and Ho) perovskites [10], RE_2FeAlO_6 (RE = Gd, Dy, Ho) oxides [11], $RE_{60}Co_{20}Ni_{20}$ (RE = Ho and Er) amorphous ribbons [12], $LaFe_{12}B_6$-based materials [13], and rare earth-based intermetallic compounds [14], were found to exhibit outstanding magnetocaloric performance at cryogenic temperatures. The development of these materials can promote the development of magnetic refrigeration technology.

$NaZn_{13}$-type $La(Fe,Si)_{13}$ compounds have attracted considerable attention as magnetocaloric materials due to their large magnetocaloric effect (MCE) values, relatively low cost, tunable Curie temperature (T_C), and environmental friendliness [1,7,15]. However, these compounds have some disadvantages [16–18]: First, it is difficult to form single-phase $La(Fe,Si)_{13}$ material. An ingot of $La(Fe,Si)_{13}$ has to be annealed above 1273 K for several days or even a month to obtain a single phase [18]. Second, the T_C of $LaFe_{13-x}Si_x$ ($1.0 \leq x \leq 1.6$) compounds is less than 210 K, which is too low for near room temperature magnetic cooling. Third, these compounds show a first-order magnetic transition (FOMT), which has a desirable large magnetic entropy change, but is accompanied by thermal and magnetic hysteresis and relatively large undesirable changes in crystal symmetry and/or volume [17]. The magneto-volume effect can lead to pulverization of bulk $LaFe_{13-x}Si_x$ magnetocaloric materials when the magnetic field is varied during the MR process [17]. The brittleness and magnetic volume change reduce formability, limiting the progress of this material for commercial applications.

Considerable research has been performed to overcome these shortcomings. Short time annealing produced a single phase material in melt-spun samples [19]. Combining kilogram-scale strip-casting with short time annealing has been deployed for mass production of $LaFe_{13-x}Si_x$-based magnetic cooling materials [20]. Elemental substitution of Fe (by Co [18,21,22], Si [23], etc.) or La (by Ce [24]) can improve magnetocaloric performance. T_C can be raised from ~200 K to room temperature or even higher by addition of interstitial atoms like H [23,25–27]. Hydrogenation of $LaFe_{13-x}Si_x$ is the most effective method to shift T_C to room temperature while maintaining a large magnetocaloric effect [28]. Quaternary $La(Fe,Co,Si)_{13}$ compounds can increase T_C and change the magnetic transition from first-order to second-order, accompanied by lower MCE.

$LaFe_{13-x}Si_x$ bulk composites can be prepared through powder metallurgy technology using $LaFe_{13-x}Si_x$ particles and binder powder as raw materials. Although $LaFe_{13-x}Si_x$ is hard and brittle, particle fracture can be avoided by adding low hardness binders during the powder metallurgy process. Additionally, the magneto-volume effect can be reduced by pulverization of the material [29]. Various binder materials, such as polymers [30], high thermal conductivity metals [31], and low melting alloys [32,33], have been studied to obtain high performance using relatively low sintering temperatures. High-temperature sintering will lead to metallurgical reactions between the binder and the particles, it will also significantly decrease the mass fraction of the 1:13 phase and lower the MCE of the composites [31]. The magnetocaloric properties of the composites arise from the La-Fe-Si based alloy. Addition of binder material reduces the mass fraction of magnetocaloric material and decreases the MCE of the composites.

La-Co alloys [34,35] have been used as binder due to the low enthalpy of formation of the $LaCo_{13}$ phase (−5.79 kJ/mol) [36], lower La-Fe binary phase content and the low melting point of a La-based eutectic alloy [37]. Based on grain boundary diffusion (GBD) theory [38], Co diffusion occurs through annealing after hot pressing [34,39]. During annealing, Co can diffuse into the 1:13 matrix, influencing the magnetic and magnetocaloric properties and promoting the formation of the 1:13 phase. The binder content and annealing parameters play an important role in the phase formation process. For example, addition of more than 5 wt.% of La-Al alloy was detrimental to 1:13 phase formation in La-Fe-Si/La-Al composites [40]. Our previous work [41] showed that excess addition of Ce-Co alloy binder would result in the formation of a new phase, which disappeared with increasing heat treatment time. However, the sintering temperature is below the melting point of $Ce_{40}Co_{60}$ (1323 K), resulting in a porous structure and insufficient mechanical properties. The large pressure of hot pressing (600 MPa) and long diffusion annealing times (up to 24 h) are unfavorable for industrial production. To overcome these limitations, we combined the advantages of hot pressing and diffusion annealing in this work. We developed a novel process to prepare La-Fe-Si-based composites by sintering to shorten the processing time and improve the compressive properties. The effect of Ce_2Co_7 binder on the formation of the 1:13 phase, thermal conductivity, mechanical, magnetic and magnetocaloric properties of $LaFe_{11.6}Si_{1.4}/Ce_2Co_7$ composites was studied. The diffusion mechanism of Ce and Co was also investigated. The T_C increased from ~200 K to near room temperature. High values of the compressive strength $(\sigma_{bc})^{max}$ of up to 450 MPa and of the thermal conductivity (λ) of up to 7.5 W/m·K were successfully obtained.

2. Experimental Section

Annealed $LaFe_{11.6}Si_{1.4}$ flakes were prepared following the procedure described in our previous work [41]. The Ce_2Co_7 compounds were prepared by an arc melting process with pure Ce ($\geq$99.5 wt.%) and Co ($\geq$99.9 wt.%). The annealed $LaFe_{11.6}Si_{1.4}$ flakes were mechanically milled to powders with a particle size in the range of 45 to 100 μm. Ce_2Co_7 fine powders were prepared with a particle size smaller than 30 μm. The $LaFe_{11.6}Si_{1.4}$ and Ce_2Co_7 powders were homogenously mixed with 0, 5, 10, and 15 wt.% content of Ce_2Co_7 powder, followed by sintering under uniaxial stress of 30 MPa at 1373 K to produce cylindrical samples with 15 mm in diameter and 5 mm in height.

XRD characterization was carried out using X-ray diffractometer (Rigaku D/max-2200/PC, Tokyo, Japan) with Cu-Kα_1 radiation (λ = 1.54056 Å) at a scan rate of 2°/min. The phase compositions and microstructures were analyzed by a thermal field emission scanning electron microscope (SEM, FEI Nova Nano SEM 430, Davis, CA, USA) equipped with an energy-dispersive spectrometer (EDS) attachment. Measurements of thermal conductivity and magnetic properties were carried out using a Physical Property Measurement System (PPMS-9, Quantum Design, San Diego, CA, USA)). The compressive strengths of the samples were measured by a mechanical testing system (Shimadzu AG-100NX, Kyoto, Japan). The porosity (P) of the composites were determined by the following equation: $P = 1 - \rho/\rho_0$, where ρ is the effective density and ρ_0 is the theoretical density. The effective density (ρ) of each sample was tested by the Archimedes method. The theoretical density (ρ_0) of each sample was calculated by the following equation: $\rho_0 = \frac{100}{\sum_i \frac{m_i}{\rho_i}}$, where m_i was the mass of the corresponding phase in the 100 g sample and the ρ_i was the density of the specific phase.

3. Results and Discussion

3.1. Phase Analysis

Figure 1a shows the Rietveld refined X-ray diffraction (XRD) patterns of $LaFe_{11.6}Si_{1.4}/x$ wt.%Ce_2Co_7 (x = 0, 5, 10, and 15) composites sintered at 1373 K for 6 h. $LaFe_{11.6}Si_{1.4}/x$ wt.% Ce_2Co_7 (x = 5, 10, and 15) sintered composites exhibited a majority 1:13 phase. The α-Fe phase content decreased, corresponding to the appearance of a $CeFe_7$-type phase (1:7 phase) and a La_2O_3 phase. For composites sintered at 1373 K for 6 h, higher Ce_2Co_7 binder content (less than 15 wt.%) can promote formation of the 1:13 phase and decrease the α-Fe phase content (Figure 1a and Table 1).

For example, $LaFe_{11.6}Si_{1.4}/5wt.\%Ce_2Co_7$ composites contained 93.21 wt.% 1:13 phase, 5.90 wt.% α-Fe phase and a small amount of 1:7 phase (0.89 wt.%). However, when the addition of Ce_2Co_7 binder rose to 10 wt.% or more, the mass fraction of 1:7 phase increased. Moreover, the mass fraction of 1:7 phase in the $LaFe_{11.6}Si_{1.4}/10$ wt.%Ce_2Co_7 composites decreased with an increase in sintering time. As reported earlier [41], the 1:7 phase occurred at grain boundaries after annealing for a short time. Clearly, the 1:7 phase content after short time annealing increased for larger binder content.

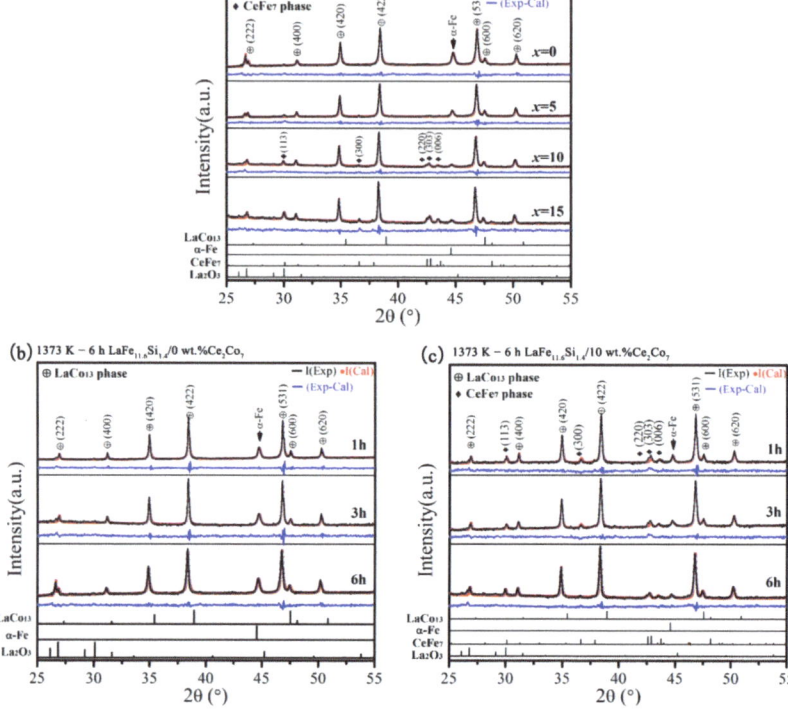

Figure 1. (a) XRD patterns for LaFe$_{11.6}$Si$_{1.4}$/$x$wt.%Ce$_2$Co$_7$ (x = 0, 5, 10 and 15) composites sintered at 1373 K for 6 h; (b) XRD patterns for binder-free LaFe$_{11.6}$Si$_{1.4}$ composites sintered at 1373 K for 1, 3 and 6 h, respectively; (c) XRD patterns for LaFe$_{11.6}$Si$_{1.4}$/10wt.%Ce$_2$Co$_7$ composites sintered at 1373 K for 1, 3 and 6 h, respectively.

Table 1. The weight percentages of phases in the LaFe$_{11.6}$Si$_{1.4}$/x wt.%Ce$_2$Co$_7$ (x = 0, 5, 10 and 15) sintered composites and fit coefficient.

Sample	Sintering@1373 K		Weight Percentage (wt.%)			Fit Coefficient			
			1:13 phase	α-Fe Phase	1:7 Phase	R_p	R_{wp}	Re_{xp}	χ^2
Annealed LaFe$_{11.6}$Si$_{1.4}$ flakes			84.69(5)	15.31(2)		0.95	1.62	0.80	4.16
LaFe$_{11.6}$Si$_{1.4}$/x wt.%Ce$_2$Co$_7$ composites	$x = 0$	1 h	89.36(5)	10.64(2)	/	0.95	1.44	1.10	1.71
		3 h	87.74(5)	12.26(2)	/	1.28	1.73	1.42	1.49
		6 h	89.26(5)	10.82(2)	/	0.87	1.36	1.01	1.80
	$x = 5$	6 h	93.21(5)	5.90(1)	0.89	0.87	1.20	1.01	1.43
	$x = 10$	1 h	87.28(5)	4.72(1)	8.00	1.08	1.46	1.26	1.33
		3 h	87.02(5)	6.28(1)	6.70	0.98	1.27	1.23	1.07
		6 h	93.93(5)	3.30(1)	2.77	0.90	1.19	0.98	1.46
	$x = 15$	6 h	88.29(5)	4.96(1)	6.75	1.19	1.53	1.09	1.98

To investigate the kinetics of diffusion mode of Ce and Co, binder-free and 10 wt.% Ce$_2$Co$_7$ composites sintered at 1373 K for 1, 3 and 6 h were prepared. The Rietveld refined. XRD patterns are shown in Figure 1b,c, respectively. The weight percentages of the phases obtained from the Rietveld refined XRD data are listed in Table 1. The phase composition of annealed LaFe$_{11.6}$Si$_{1.4}$ flakes was 84.69(5) wt.% 1:13 phase, and 15.31(2) wt.% α-Fe phase. There were no peaks of the La$_2$O$_3$, LaFeSi (1:1:1), and (La,Ce)$_2$(Fe,Co,Si)$_{17}$ (2:17) phases

(Figure 1) in the XRD patterns due to the limitation of the resolution of normal XRD, indicating the low content of these phases in the sintered composites.

Compared with the annealed flakes, the 1:13 phase content in the binder-free sample sintered at 1373 K for 1~6 h (Figure 1b) slightly increased, accompanied by a decrease in α-Fe phase content. This indicated that additional formation of the 1:13 phase occurred during sintering. The XRD patterns of LaFe$_{11.6}$Si$_{1.4}$/10 wt.%Ce$_2$Co$_7$ composites sintered at 1373 K for 1, 3 and 6 h are displayed in Figure 1c. When the sintering time was extended from 1 h to 6 h, the 1:7 phase content reduced from 8.00 wt.% to 2.77 wt.% (Table 1), indicating that increased high-temperature sintering time can significantly reduce the 1:7 phase content.

3.2. Microstructure Evolution

The microstructure of the composites sintered at 1373 K for 6 h is shown in Figure 2 for various values of Ce$_2$Co$_7$ binder content. For the binder-free sample, the black area is the α-Fe phase, the dark gray area is the 1:13 phase (matrix phase), and white grains consist of the La$_2$O$_3$ phase (not detected in the XRD results owing to its low mass fraction). After sintering at 1373 K for 6 h, a high mass fraction of the 1:13 phase and a small α-Fe phase content (Table 1) were obtained. It was reported that the temperature range for thermal decomposition (TD) reaction for La(Fe,Si,Co)$_{13}$ based compounds is 873–1173 K [42]. A lamellar structure, usually observed upon decomposition of the 1:13 phase was not seen, indicating no decomposition of the 1:13 phase occurred during sintering of the binder-free LaFe$_{11.6}$Si$_{1.4}$ composites [43]. The sintering temperature was near the formation temperature of the 1:13 phase, thus, a large amount of precipitation was observed (Figure 2b). The α-Fe phase (black areas), the LaFeSi (1:1:1) phase (white grains inside the 1:13 particles) and the La$_2$O$_3$ phase (white grains besides the α-Fe phase), shown in Figure 2b, were observed. As previously reported [42,43], the white phase at the particle interfaces (Figure 2a) is believed to be a La- and Si-rich phase.

From the Ce-Co binary phase diagram [44], the melting point of Ce$_2$Co$_7$ alloy is 1418 K, which is 45 K higher than the sintering temperature (1373 K). Therefore, the soft Ce$_2$Co$_7$ alloy binder and LaFe$_{11.6}$Si$_{1.4}$ particles could be bonded by solid phase sintering to obtain a highly dense microstructure under a pressure of 30 MPa [45,46].

For the LaFe$_{11.6}$Si$_{1.4}$/5 wt.%Ce$_2$Co$_7$ sintered composites (Figure 2c,d), the fraction of α-Fe phase (black area), La- and Si-rich boundary phase and La$_2$O$_3$ phase (white area) decreased markedly. A small amount of CeFe$_7$-type phase (light grey) appeared in the binder bonded composites. The formation enthalpy of CeCo$_7$ (−10.704 kJ/mol calculated by the modified Miedema (ZSL's Model) [36] is more negative than that of LaCo$_{13}$ (−5.79 kJ/mol) [34], favoring the formation of the 1:7 phase. The EDS map results of the LaFe$_{11.6}$Si$_{1.4}$/5 wt.%Ce$_2$Co$_7$ sintered composites are shown in the inset of Figure 2c, uniform diffusion of Co occurred in the 1:13 phase. However, there are some regions of La and Ce enrichment, accompanied by up to ~68 at.% for oxygen (Figure 2d). Owing to the limitation of EDS in quantifying the concentration of light elements, this value for oxygen must be regarded as a rough estimate. These La- and Ce-rich regions can correspond to the (La,Ce)$_2$O$_3$ phase. Ce is mainly found in the 1:7 phase, a small amount of Ce is found in the 1:13 phase. Sintering at 1373 K for 6 h cannot produce a uniform distribution of Ce or eliminate the 1:7 phase.

Interestingly, the LaFeSi (1:1:1), La-, and Si-rich phases are virtually absent in LaFe$_{11.6}$Si$_{1.4}$/x wt.%Ce$_2$Co$_7$ (x = 5, 10, 15) composites sintered at 1373 K for 6 h. The chemical compositions of the 1:7 and 1:13 phases and lattice constant (a) of 1:13 phase for the sintered composites are listed in Table 2. The 1:13 phase, with a composition of La$_{0.87}$Ce$_{0.13}$Fe$_{9.70}$Co$_{0.56}$Si$_{1.35}$, La$_{0.82}$Ce$_{0.18}$Fe$_{9.24}$Co$_{0.91}$Si$_{1.34}$, and La$_{0.75}$Ce$_{0.25}$Fe$_{9.10}$Co$_{1.30}$Si$_{1.28}$ was obtained in sintered composites with 5 wt.%, 10 wt.%, and 15 wt.% Ce$_2$Co$_7$, respectively.

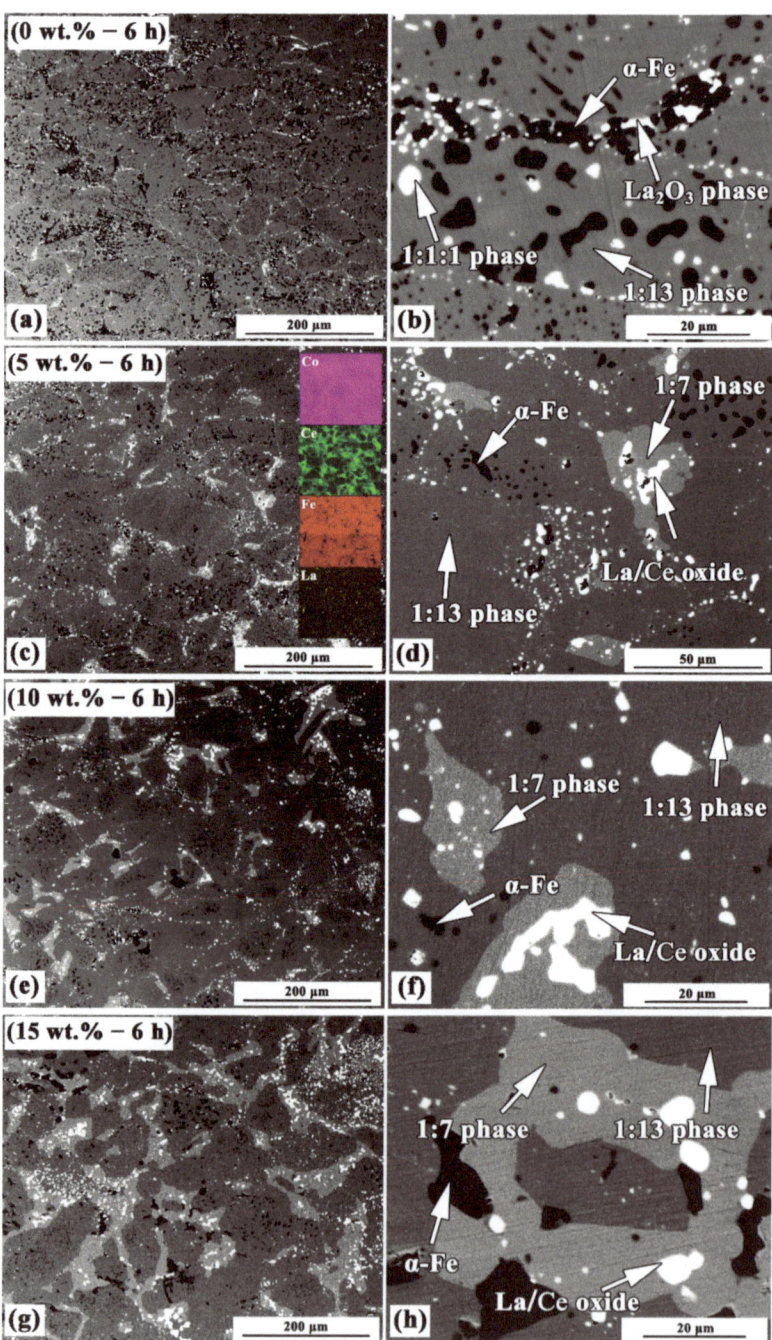

Figure 2. Backscattered SEM micrographs of LaFe$_{11.6}$Si$_{1.4}$/x wt.%Ce$_2$Co$_7$ composites with $x = 0$ (**a,b**), 5 (**c,d**), 10 (**e,f**), and 15 (**g,h**) sintered at 1373 K for 6 h, respectively. The inset of (**c**) shows the mapping results of La, Ce, Fe and Co elements for LaFe$_{11.6}$Si$_{1.4}$/5 wt.%Ce$_2$Co$_7$ composites sintered at 1373 K for 6 h, respectively.

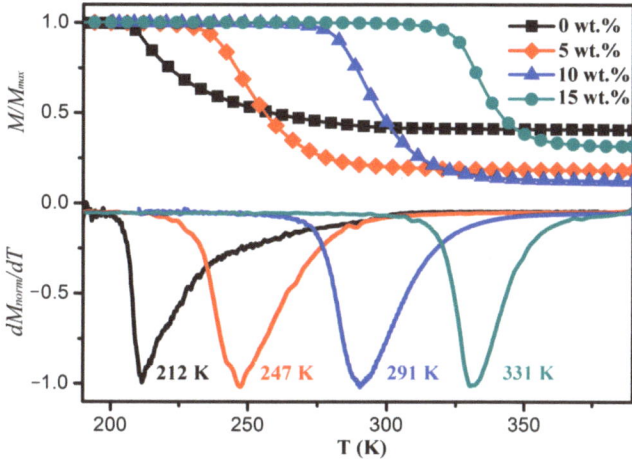

Figure 5. Normalized M/M_{max}–T and dM_{norm}/dT–T curves of LaFe$_{11.6}$Si$_{1.4}/x$ wt.%Ce$_2$Co$_7$ ($x = 0, 5, 10,$ and 15) composites sintered at 1373 K for 6 h upon cooling under an applied field of 0.05 T.

Another parameter called RCP (Relative Cooling Power) can be employed to evaluate the refrigeration efficiency. RCP was defined as the product of $(-\Delta S_M)^{max}$ times the full temperature width at half maximum of the peak (RCR = $(-\Delta S_M)^{max} \times \delta T_{FWHM}$) [54]. The values of T_C, $(-\Delta S_M)^{max}$ and RCP of the studied composites are presented in Table 2.

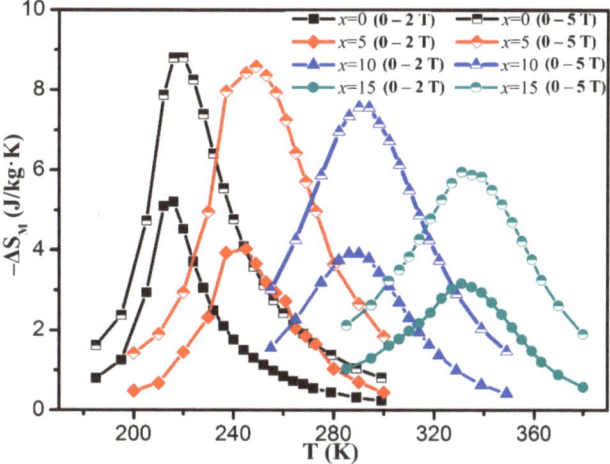

Figure 6. Magnetic entropy change $(-\Delta S_M)$ as function of temperature for the LaFe$_{11.6}$Si$_{1.4}/x$ wt.%Ce$_2$Co$_7$ ($x = 0, 5, 10$ and 15) composites sintered at 1373 K for 6 h.

The binder-free sintered composites show higher T_C (212 K) than that of the annealed LaFe$_{11.6}$Si$_{1.4}$ flakes (196 K). The sintering temperature is close to the optimum formation temperature of the 1:13 phase [55], which may bring about 1:13 phase formation and decrease the α-Fe phase content (Table 1), resulting in the formation of a nonstoichiometric Fe deficient 1:13 phase (Table 2). This Fe-deficient 1:13 phase has elevated T_C (Table 2), which weakens the first order magnetic transition (FOMT) and lowers the $(-\Delta S_M)^{max}$ [7,18,34]. The binder-free composites exhibited the maximum magnetic entropy change $(-\Delta S_M)^{max}$ of 5.2 and 8.8 J/kg·K under applied field changes of 0–2 and 0–5 T, respectively (Table 2).

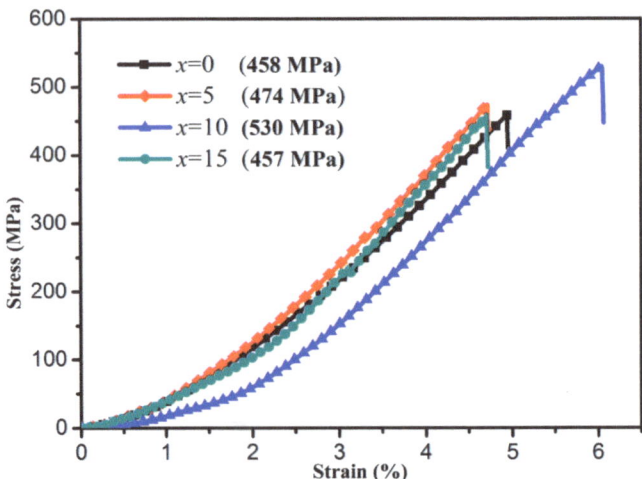

Figure 4. Stress–strain curves for the LaFe$_{11.6}$Si$_{1.4}$/x wt.%Ce$_2$Co$_7$ (x = 0, 5, 10 and 15) composites sintered at 1373 K for 6 h.

Table 3. Porosity, thermal conductivity, and mechanical properties of LaFe$_{11.6}$Si$_{1.4}$/x wt.%Ce$_2$Co$_7$ (x = 0, 5, 10 and 15) composites sintered at 1373 K for 6 h.

Sample	Porosity (%)	λ (W/m·K)	$(\sigma_{bc})^{max}$ (MPa)	Strain (%)
Binder free	5.14	13.71	458	5.0
5 wt.%	4.02	7.81	474	4.7
10 wt.%	4.78	8.21	530	6.1
15 wt.%	4.56	7.50	457	4.7

The thermal conductivity of these composites are comparable or larger than the values for hot-pressed La$_{0.7}$Ce$_{0.3}$Fe$_{11.45}$Mn$_{0.15}$Si$_{1.4}$/13.5%Fe (7.5 W/m·K) [51] and LaFe$_{11.6}$Si$_{1.4}$H$_y$/Sn (7.9 W/m·K) [32], but smaller than those of hot-pressed La$_{0.8}$Ce$_{0.2}$(Fe$_{0.95}$Co$_{0.05}$)$_{11.8}$Si$_{1.2}$/Sn$_{42}$Bi$_{58}$ (10.72–19.64 W/m·K) [33], LaFe$_{11}$Co$_{0.8}$Si$_{1.2}$/Al (9.9~17.0 W/m·K) [52], and La$_{0.7}$Ce$_{0.3}$Fe$_{11.48}$Mn$_{0.12}$Si$_{1.4}$H$_{1.8}$/20 wt.%In (11.5 W/m·K) [53]. As La(Fe,Si)$_{13}$ and residual Ce-Co phase possess different thermal conductivities, thermal conductivity of sintered composites may be improved by even longer sintering time. The strain values are larger than 4.7%, which is about 4.5 times the values of sintered LaFe$_{11.7}$Co$_{1.3}$Si composites [42]. Thus, the ductility of the LaFe$_{11.6}$Si$_{1.4}$/x wt.%Ce$_2$Co$_7$ (x = 0, 5, 10 and 15) composites sintered at 1373 K for 6 h is markedly improved. These improved properties are mainly due to the stress distribution in the α-Fe and 1:7 phases and the improved quality of the sintered compacts during high temperature sintering. High ultimate compressive stress and better elongation can lead to better machinability.

The normalized M_{norm}–T (M/M_{max}–T) curves upon cooling process under 0.05 T, normalized magnetization temperature derivative curves (dM_{norm}/dT–T) measured during cooling under an applied field of 0.05 T and the $(-\Delta S_M)$–T curves measured under applied field changes of 0–2 and 0–5 T for the LaFe$_{11.6}$Si$_{1.4}$/x wt.%Ce$_2$Co$_7$ (x = 0, 5, 10 and 15) composites sintered at 1373 K for 6 h are shown in Figures 5 and 6, respectively. The isothermal magnetization curves for the LaFe$_{11.6}$Si$_{1.4}$/x wt.%Ce$_2$Co$_7$ were measured with an increasing magnetic field in a wide temperature range. The sweep rate of the field was slow enough to ensure that the data were recorded in an isothermal process. The value of the isothermal magnetic entropy change $-\Delta S_M(T, H)$ is given by the Maxwell relationship:

$$-\Delta S_M(T,H) = \int_0^H (\partial M/\partial T) dH \qquad (2)$$

The T_C of the LaFe$_{11.6}$Si$_{1.4}$/x wt.%Ce$_2$Co$_7$ (x = 5, 10, and 15) composites increased linearly with larger Ce$_2$Co$_7$ content due to the higher Co content diffused into the 1:13 phase. Sintering caused Ce and Co to diffuse from the Ce$_2$Co$_7$ binder to the 1:13 particles through the particle boundaries. It changed the chemical composition of the 1:13 phase, the 1:13 phase content, and the magnetocaloric performance. As calculated from the results in Table 2, the content of Ce and Co in the 1:13 phase increased from 1.03 to 1.97 at.% and 4.44 to 10.25 at.%, respectively. The T_C of sintered composites increased from 247 K to 331 K due to the substitution of Fe by Co in the 1:13 phase. Under the applied field changes of 2 T and 5 T, the $(-\Delta S_M)^{max}$ values of the LaFe$_{11.6}$Si$_{1.4}$/x wt.%Ce$_2$Co$_7$ composites with x = 0, 5, 10 and 15 were 5.2, 4.0, 3.9, 3.2 J/kg·K and 8.8, 8.6, 7.6, 6.0 J/kg·K, respectively. RCP values of these composites were 146.7, 158.4, 209.2, and 178.3 J/kg and 339.0, 428.7, 457.7, and 399.2 J/kg, respectively (Table 2).

Interestingly, the $(-\Delta S_M)^{max}$ (~5.2 J/kg·K@2 T at 212 K) of the binder-free composite was larger than that of the LaFe$_{11.6}$Si$_{1.4}$ bulk composite with particle size smaller than 100 μm (~4.0 J/kg·K@2 T at 220 K) [56], and far larger than that of La(Fe,Co)$_{11.4}$Al$_{1.6}$ alloy (2.2 J/kg·K@2 T at 205 K) [50] in a similar temperature range. The $(-\Delta S_M)^{max}$ value (~4 J/kg·K@2 T at 247 K) of the composite with 5 wt.% Ce$_2$Co$_7$ binder was larger than those of LaFe$_{11.6}$Si$_{1.4}$/5 wt.%Pr$_{40}$Co$_{60}$ (~2.9 J/kg·K@2 T at 247 K) [56] and Gd$_{50}$Co$_{48}$Zn$_2$ as-spun ribbons [57] (2.6 J/kg·K@2 T at 260 K) in a similar temperature range. Compared with Gd [58], the values of $(-\Delta S_M)^{max}$ (~3.9 J/kg·K@2 T at 291 K) and RCP (209.2 J/kg@2 T) of the LaFe$_{11.6}$Si$_{1.4}$/10 wt.%Ce$_2$Co$_7$ composite is smaller. For an applied field change of 2 T, the values of $(-\Delta S_M)^{max}$ and RCP for the LaFe$_{11.6}$Si$_{1.4}$/15 wt.%Ce$_2$Co$_7$ composite (Table 2) are much larger than those of Fe$_{77}$Ta$_3$B$_{10}$Zr$_9$Cu$_1$ amorphous ribbons (($-\Delta S_M)^{max}$~1.47 J/kg·K@2 T at 336 K; RCP~123.9 J/kg) in the similar temperature [59].

The Arrott plots near T_C of the LaFe$_{11.6}$Si$_{1.4}$/x wt.%Ce$_2$Co$_7$ (x = 0, 5, 10 and 15) composites sintered at 1373 K for 6 h are shown in Figure 7. According to the Banerjee criterion [60], the slopes of the Arrott plots for the composites sintered at 1373 K for 6 h are all positive. The magnetic phase transition of these sintered composites near its T_C is a second-order magnetic phase transition, which is due to Co diffusion in Ce$_2$Co$_7$ bonded composites and enrichment of Si in the 1:13 matrix for the binder-free composite. A similar behavior was observed for the binder-free and 10 wt.% Ce$_2$Co$_7$ binder composites sintered at 1373 K for 1~6 h, which also exhibited a second-order magnetic phase transition near their T_C.

The normalized M/M_{max}–T curves upon cooling under an applied field of 0.05 T of the binder-free and 10 wt.% Ce$_2$Co$_7$ binder composites sintered at 1373 K are shown in Figure 8. The $(-\Delta S_M)$–T curves (0 − 2 T) are shown in the inset of Figure 8, respectively. For the binder-free composites, there is a composition change of the magnetocaloric phase (1:13 phase) after sintering. Based on earlier work [61], the decrease of Fe/Si ratio in 1:13 phase results in higher T_C and lower $-\Delta S_M$. With increasing annealing time from 1 to 6 h, the Fe/Si ratio in the binder-free composites decreased from ~6.89 (10.95/1.59) to ~6.58 (10.53/1.60), the T_C increased from ~206 to 212 K while $(-\Delta S_M)^{max}$ decreased from 7.5 to 5.2 J/kg·K (ΔH = 2 T) (Table 2). The 10 wt.% Ce$_2$Co$_7$ binder composites also exhibited higher $(-\Delta S_M)^{max}$ due to the diffusion of Ce, which increased the Ce content in the 1:13 phase. Hence, the change of T_C and magnetocaloric properties is attributed to the formation of the 1:13 phase (Table 1) and grain boundary diffusion of Ce and Co atoms.

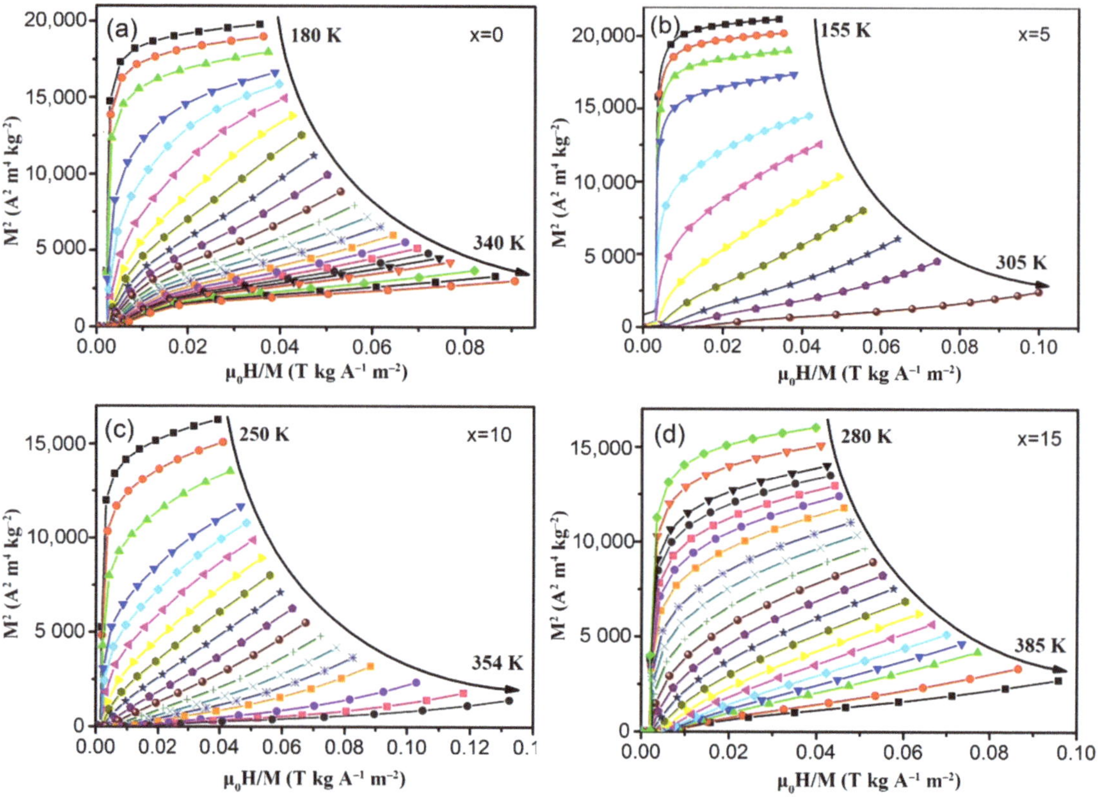

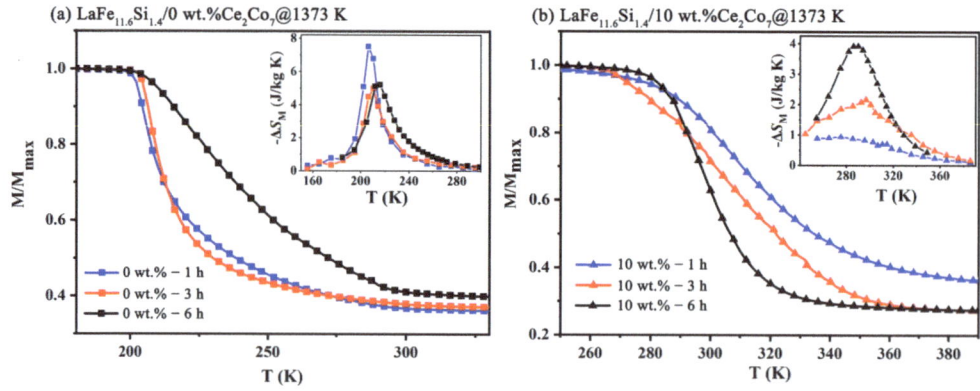

Figure 7. Arrott plots (**a**–**d**) of the LaFe$_{11.6}$Si$_{1.4}$/$x$wt.%Ce$_2$Co$_7$ (x = 0, 5, 10, and 15) composites sintered at 1373 K for 6 h.

Figure 8. Normalized M/M_{max}–T curves of the binder-free and 10 wt.% Ce$_2$Co$_7$ binder composites sintered at 1373 K for 1~6 h upon cooling under an applied field of 0.05 T. Insets in panels (**a**,**b**) are ($-\Delta S_M$) vs. T curves for the composites with and without Ce$_2$Co$_7$ binder under a magnetic field change of 0–2 T, respectively.

4. Conclusions

In summary, the diffusion of Ce and Co into the desired $(La,Ce)(Fe,Co,Si)_{13}$ phase and the partial substitution of La by Ce and Fe by Co could be realized by a one-step sintering of $La(Fe,Si)_{13}$/Ce-Co composites. The line scan of the interphase boundary between the 1:13 phase and the 1:7 phase showed a sudden change in Ce content and little change in Co concentration. Thus, the diffusion mechanism of Ce in the 1:13 phase was different from that of Co. Co diffused by the usual vacancy diffusion mechanism while the diffusion mode of Ce was by reaction diffusion. After 6 h sintering at 1373 K, a high value of compressive strength $(\sigma_{bc})^{max}$ of up to 450 MPa and a high thermal conductivity (λ) of up to 7.5 W/m·K were successfully obtained. The magnetocaloric properties of the composites could also be tuned by varying the content of Ce-Co alloy binder and the sintering time. T_C increased from 212 K to 331 K with increasing Ce_2Co_7 from 0 to 15 wt.% in $LaFe_{11.6}Si_{1.4}$/Ce-Co bulk composites. The magnetic entropy change decreased from 8.8 to 6.0 J/kg·K, the relative cooling power increased from ~339 to 428, 457 J/kg and then decreases to 399 J/kg under 5 T.

Thus, a feasible route to produce high quality $La(Fe,Si)_{13}$ based magnetocaloric composites with large MCE, good mechanical properties and thermal conductivity and tunable T_C by a one-step sintering process has been demonstrated.

Author Contributions: Conceptualization, Methodology, Supervision, Writing—original draft, X.-C.Z.; Conceptualization, Methodology, Formal analysis, Investigation, Data curation, Writing—original draft, X.-T.D.; Software, Resources, Validation, J.-H.H., C.-L.L., H.Z. and Y.-L.H. Visualization, Investigation, Formal analysis, H.-Y.Y.; Writing—review & editing R.V.R. All authors have read and agreed to the published version of the manuscript.

Funding: This work was funded by the National Natural Science Foundation of China (Grant Nos. 51874143, 52066001, 51671022, 51461012), the Natural Science Foundation of Guangdong Province (Grant Nos. 2019A1515010970, 2017A030313317), the National Key Research & Development Program of China (Materials Genome Initiative) (Grant No. 2017YFB0702703), the Guangzhou Municipal Science and Technology Project (Grant No. 201904010030), and the Foundation of State Key Laboratory of Baiyunobo Rare Earth Resource Researches and Comprehensive Utilization (Grant No. 2020Z2218). This Research is also supported by Singapore-HUJ Alliance for Research and Enterprise (SHARE), Nanomaterials for Energy and Energy-Water Nexus (NEW), Campus for Research Excellence and Technological Enterprise (CREATE), Singapore 138602. And the APC was funded by the National Natural Science Foundation of China (Grant No. 51874143).

Institutional Review Board Statement: Not applicable.

Informed Consent Statement: Not applicable.

Data Availability Statement: The data used to support the findings of this study are available from the corresponding author upon request.

Acknowledgments: The authors are grateful to Chun-Ming Li for experimental assistance. The authors wish to also thank Yu-Cai Wu for modifying the figures of this paper.

Conflicts of Interest: The authors declare that they have no known competing financial interests or personal relationships that could have appeared to influence the work reported in this paper.

References

1. Gutfleisch, O.; Willard, M.A.; Brück, E.; Chen, C.H.; Sankar, S.G.; Liu, J.P. Magnetic Materials and Devices for the 21st Century: Stronger, Lighter, and More Energy Efficient. *Adv. Mater.* **2011**, *23*, 821–842. [CrossRef] [PubMed]
2. Chaudhary, V.; Chen, X.; Ramanujan, R.V. Iron and manganese based magnetocaloric materials for near room temperature thermal management. *Prog. Mater. Sci.* **2019**, *100*, 64–98. [CrossRef]
3. Pecharsky, V.K.; Gschneidner, K.A., Jr. Giant Magnetocaloric Effect in $Gd_5(Si_2Ge_2)$. *Phys. Rev. Lett.* **1997**, *78*, 4494–4497. [CrossRef]
4. Virgil, P.; Alexander, J.S.; Robert, D.S. Reduction of hysteresis losses in the magnetic refrigerant $Gd_5Ge_2Si_2$ by the addition of iron. *Nature* **2004**, *429*, 853–857.
5. Tegus, O.; Brück, E.; Buschow, K.H.J.; De Boer, F.R. Transition-metal-based magnetic refrigerants for room-temperature applications. *Nature* **2002**, *415*, 150–152. [CrossRef] [PubMed]

6. Li, Y.; Zeng, Q.Q.; Wei, Z.Y.; Liu, E.K.; Han, X.L.; Du, Z.W.; Li, L.W.; Xi, X.K.; Wang, W.H.; Wang, S.G.; et al. An efficient scheme to tailor the magnetostructural transitions by staged quenching and cyclical ageing in hexagonal martensitic alloys. *Acta Mater.* **2019**, *174*, 289–299. [CrossRef]
7. Hu, F.X.; Shen, B.G.; Sun, J.R.; Cheng, Z.H.; Rao, G.H.; Zhang, X.X. Influence of negative lattice expansion and meta-magnetic transition on magnetic entropy change in the compound LaFe$_{11.4}$Si$_{1.6}$. *Appl. Phys. Lett.* **2001**, *78*, 3675–3677. [CrossRef]
8. Fujita, A.; Fujieda, S.; Fukamichi, K.; Mitamura, H.; Goto, T. Itinerant-electron metamagnetic transition and large magne-tovolume effects in La(Fe$_x$Si$_{1-x}$)$_{13}$ compounds. *Phys. Rev.* **2001**, *B65*, 014410. [CrossRef]
9. Krenke, T.; Duman, E.; Acet, M.; Wassermann, E.F.; Moya, X.; Mañosa, L.; Planes, A. Inverse magnetocaloric effect in ferromagnetic Ni-Mn-Sn alloys. *Nat. Mater.* **2005**, *4*, 450–454. [CrossRef]
10. Li, L.W.; Xu, P.; Ye, S.K.; Li, Y.; Liu, G.D.; Huo, D.X.; Yan, M. Magnetic properties and excellent cryogenic magnetocaloric performances in B-site ordered RE$_2$ZnMnO$_6$ (RE=Gd, Dy and Ho) perovskites. *Acta Mater.* **2020**, *194*, 354–365. [CrossRef]
11. Wu, B.B.; Zhang, Y.K.; Guo, D.; Wang, J.; Ren, Z.M. Structure, magnetic properties and cryogenic magneto-caloric effect (MCE) in RE$_2$FeAlO$_6$ (RE=Gd, Dy, Ho) oxides. *Ceram. Int.* **2021**, *47*, 6290–6297. [CrossRef]
12. Wang, Y.M.; Guo, D.; Wu, B.B.; Geng, S.H.; Zhang, Y.K. Magnetocaloric effect and refrigeration performance in RE$_{60}$Co$_{20}$Ni$_{20}$ (RE = Ho and Er) amorphous ribbons. *J. Magn. Magn. Mater.* **2019**, *498*, 166179. [CrossRef]
13. Ma, Z.P.; Dong, X.S.; Zhang, Z.Q.; Li, L.W. Achievement of promising cryogenic magnetocaloric performances in La$_{1-x}$Pr$_x$Fe$_{12}$B$_6$ compounds. *J. Mater. Sci. Technol.* **2021**, *92*, 138–142. [CrossRef]
14. Li, L.W.; Yan, M. Recent progresses in exploring the rare earth based intermetallic compounds for cryogenic magnetic refrigeration. *J. Alloys Compd.* **2020**, *823*, 153810. [CrossRef]
15. Zou, J.D.; Shen, B.G.; Gao, B.; Shen, J.; Sun, J.R. The magnetocaloric effect of LaFe$_{11.6}$Si$_{1.4}$, La$_{0.8}$Nd$_{0.2}$Fe$_{11.5}$Si$_{1.5}$, and Ni$_{43}$Mn$_{46}$Sn$_{11}$ compounds in the vicinity of the first-order phase transition. *Adv. Mater.* **2009**, *21*, 693–696. [CrossRef]
16. Shen, B.G.; Sun, J.R.; Hu, F.X.; Zhang, H.W.; Cheng, Z.H. Recent Progress in Exploring Magnetocaloric Materials. *Adv. Mater.* **2009**, *21*, 4545–4564. [CrossRef]
17. Lyubina, J.; Schafer, R.; Martin, N.; Schultz, L.; Gutfleisch, O. Novel design of La(Fe,Si)$_{13}$ alloys towards high magnetic refrigeration performance. *Adv. Mater.* **2010**, *22*, 3735–3739. [CrossRef] [PubMed]
18. Hu, F.X.; Shen, B.G.; Sun, J.R.; Zhang, X.X. Great magnetic entropy change in La(Fe,M)$_{13}$ (M=Si, Al) with Co doping. *Chin. Phys.* **2000**, *9*, 550–556.
19. Gutfleisch, O.; Yan, A.; Müller, K.H. Large magnetocaloric effect in melt-spun LaFe$_{13-x}$Si$_x$. *J. Appl. Phys.* **2005**, *97*, 10M305. [CrossRef]
20. Zhong, X.C.; Feng, X.L.; Huang, J.H.; Zhang, H.; Huang, Y.L.; Liu, Z.W.; Jiao, D.L. Microstructure evolution and large magnetocaloric effect of La$_{0.8}$Ce$_{0.2}$(Fe$_{0.95}$Co$_{0.05}$)$_{11.8}$Si$_{1.2}$ alloy prepared by strip-casting and annealing. *Aip Adv.* **2018**, *8*, 048102.
21. Liu, X.B.; Altounian, Z. Effect of Co content on magnetic entropy change and structure of La(Fe$_{1-x}$Co$_x$)$_{11.4}$Si$_{1.6}$. *J. Magn. Magn. Mater.* **2003**, *264*, 209–213. [CrossRef]
22. Hu, F.X.; Gao, J.; Qian, X.L.; Ilyn, M.; Tishin, A.M.; Sun, J.R.; Shen, B.G. Magnetocaloric effect in itinerant electron metamagnetic systems La(Fe$_{1-x}$Co$_x$)$_{11.9}$Si$_{1.1}$. *J. Appl. Phys.* **2005**, *97*, 10M303. [CrossRef]
23. Fujita, A.; Fujieda, S.; Hasegawa, Y.; Fukamichi, K. Itinerant-electron metamagnetic transition and large magnetocaloric effects in La(Fe$_x$Si$_{1-x}$)$_{13}$ compounds and their hydrides. *Phys. Rev.* **2003**, *B67*, 104416. [CrossRef]
24. Fujieda, S.; Fujita, A.; Fukamichi, K. Enhancements of magnetocaloric effects in La(Fe$_{0.90}$Si$_{0.10}$)$_{13}$ and Its hydride by partial substitution of Ce for La. *Mater. Trans.* **2004**, *45*, 3228–3231. [CrossRef]
25. Mandal, K.; Pal, D.; Gutfleisch, O.; Kerschl, P.; Müller, K.H. Magnetocaloric effect in reactively-milled LaFe$_{11.57}$Si$_{1.43}$H$_y$ intermetallic compounds. *J. Appl. Phys.* **2007**, *102*, 053906. [CrossRef]
26. Krautz, M.; Moore, J.D.; Skokov, K.P.; Liu, J.; Teixeira, C.S.; Schäfer, R.; Schultz, L.; Gutfleisch, O. Reversible solid-state hydrogen-pump driven by magnetostructural transformation in the prototype system La(Fe,Si)$_{13}$H$_y$. *J. Appl. Phys.* **2012**, *112*, 083918. [CrossRef]
27. Lyubina, J.; Hannemann, U.; Ryan, M.P.; Cohen, L.F. Electrolytic hydriding of LaFe$_{13-x}$Si$_x$ alloys for energy efficient magnetic cooling. *Adv. Mater.* **2012**, *24*, 2042–2046. [CrossRef] [PubMed]
28. Zheng, H.; Tang, Y.; Chen, Y.; Wu, J.; Wang, H.; Xue, X.; Wang, J.; Pang, W. The high-temperature hydrogenation behavior of LaFe$_{11.6}$Si$_{1.4}$ and splitting of LaFe$_{11.6}$Si$_{1.4}$H$_y$ magnetocaloric transition. *J. Alloys Compd.* **2015**, *646*, 124–128. [CrossRef]
29. Hu, F.X.; Chen, L.; Wang, J.; Bao, L.F.; Sun, J.R.; Shen, B.G. Particle size dependent hysteresis loss in La$_{0.7}$Ce$_{0.3}$Fe$_{11.6}$Si$_{1.4}$C$_{0.2}$ first-order systems. *Appl. Phys. Lett.* **2012**, *100*, 072403. [CrossRef]
30. Skokov, K.; Karpenkov, D.; Kuz'Min, M.D.; Radulov, I.A.; Gottschall, T.; Kaeswurm, B.; Fries, M.; Gutfleisch, O. Heat exchangers made of polymer-bonded La(Fe,Si)$_{13}$. *J. Appl. Phys.* **2014**, *115*, 17A941. [CrossRef]
31. Liu, J.; Zhang, M.X.; Shao, Y.Y.; Yan, A.R. LaFe11.6Si1.4/Cu magnetocaloric composites prepared by hot pressing. *IEEE Trans. Magn.* **2015**, *51*, 2501502.
32. Zhang, H.; Liu, J.; Zhang, M.; Shao, Y.; Li, Y.; Yan, A. LaFe$_{11.6}$Si$_{1.4}$H$_y$/Sn magnetocaloric composites by hot pressing. *Scr. Mater.* **2016**, *120*, 58–61. [CrossRef]
33. Dong, X.T.; Zhong, X.C.; Peng, D.R.; Huang, J.H.; Zhang, H.; Jiao, D.L.; Liu, Z.W.; Ramanujan, R.V. La$_{0.8}$Ce$_{0.2}$(Fe$_{0.95}$Co$_{0.05}$)$_{11.8}$Si$_{1.2}$/Sn$_{42}$Bi$_{58}$ magnetocaloric composites prepared by low temperature hot pressing. *J. Alloys Compd.* **2018**, *737*, 568–574. [CrossRef]

34. Zhong, X.C.; Peng, D.R.; Dong, X.T.; Huang, J.H.; Zhang, H.; Jiao, D.L.; Liu, Z.W.; Ramanujan, R.V. Improvement in the magnetocaloric properties of sintered La(Fe,Si)13 based composites processed by La-Co grain boundary diffusion. *J. Alloys Compd.* **2019**, *780*, 873–880. [CrossRef]
35. Zhong, X.C.; Peng, D.R.; Dong, X.T.; Huang, J.H.; Zhang, H.; Huang, Y.L.; Wu, S.M.; Yu, H.Y.; Qiu, W.Q.; Liu, Z.W.; et al. Improvement in mechanical and magnetocaloric properties of hot-pressed La(Fe,Si)13/La70Co30 composites by grain boundary engineering. *Mater. Sci. Eng. B* **2021**, *263*, 114900. [CrossRef]
36. Zhang, R.F.; Zhang, S.H.; He, Z.J.; Jing, J.; Sheng, S.H. Miedema Calculator: A thermodynamic platform for predicting formation enthalpies of alloys within framework of Miedema's Theory. *Comput. Phys. Commun.* **2016**, *209*, 58–69. [CrossRef]
37. Guo, Q.W.; Wang, G.S.; Guo, G.C. *Non-Ferrous Metal Atlas of Binary Alloy Phase*; Chemical Industry Press: Beijing, China, 2010.
38. Zhou, Q.; Liu, Z.W.; Zhong, X.C.; Zhang, G.Q. Properties improvement and structural optimization of sintered NdFeB magnets by non-rare earth compound grain boundary diffusion. *Mater. Des.* **2015**, *86*, 114–120. [CrossRef]
39. Chen, X.; Chen, Y.G.; Tang, Y.B. The effect of high-temperature annealing on LaFe$_{11.5}$Si$_{1.5}$ and the magnetocaloric properties of La$_{1-x}$Ce$_x$Fe$_{11.5}$Si$_{1.5}$ compounds. *Rare Metals* **2011**, *30*, 343–347. [CrossRef]
40. Fan, W.B.; Hou, Y.H.; Ge, X.J.; Huang, Y.L.; Luo, J.M.; Zhong, Z.C. Microstructure and improved magnetocaloric properties: LaFeSi/LaAl magnets prepared by spark plasma sintering technique. *J. Phys. D Appl. Phys.* **2018**, *51*, 115003. [CrossRef]
41. Zhong, X.C.; Dong, X.T.; Peng, D.R.; Huang, J.H.; Zhang, H.; Jiao, D.L.; Zhang, H.; Liu, Z.W.; Ramanujan, R.V. Table-like magnetocaloric effect and enhanced refrigerant capacity of HPS La(Fe,Si)13 based composites by Ce-Co grain boundary diffusion. *J. Mater. Sci.* **2020**, *55*, 5908–5919. [CrossRef]
42. Löwe, K.; Liu, J.; Skokov, K.; Moore, J.D.; Amin, H.S.; Hono, K.; Katter, M.; Gutfleisch, O. The effect of the thermal decom-position reaction on the mechanical and magnetocaloric properties of La(Fe,Si,Co)$_{13}$. *Acta Mater.* **2012**, *60*, 4268–4276. [CrossRef]
43. Liu, J.; Krautz, M.; Skokov, K.; Woodcock, T.G.; Gutfleisch, O. Systematic study of the microstructure, entropy change and adiabatic temperature change in optimized La-Fe-Si alloys. *Acta Mater.* **2011**, *59*, 3602–3611. [CrossRef]
44. Predel, B. Ce-Co (Cerium-Cobalt) phase diagram. In *Phase Equilibria, Crystallographic and Thermodynamic Data of Binary Alloys Landolt-Börnstein: Numerical Data and Functional Relationships in Science and Technology-New Series*; Springer: Berlin/Heidelberg, Germany, 1993; Volume 3, ISBN 9783540614333.
45. Laptev, A.V. Structure and properties of WC-Co alloys in solid-phase sintering. I. Geometrical evolution. *Powder Met. Met. Ceram.* **2007**, *46*, 415–422. [CrossRef]
46. Pribytkov, G.A.; Andreeva, I.A.; Korzhova, V.V. Bulk changes and structure formation in solid-phase sintering of Ti − TiAl$_3$ powder mixtures. *Powder Met. Met. Ceram.* **2008**, *47*, 687–692. [CrossRef]
47. Itoh, M.; Machida, K.I.; Hirose, K.; Adachi, G.Y. Nitrogen absorption and desorption characteristics for CeFe$_7$. *Chem. Lett.* **2001**, *30*, 294–295. [CrossRef]
48. Hou, X.L.; Lampen-Kelley, P.; Xue, Y.; Liu, C.Y.; Xu, H.; Han, N.; Ma, C.W.; Srikanth, H.; Phan, M.H. Formation mechanisms of NaZn$_{13}$-type phase in giant magnetocaloric La–Fe–Si compounds during rapid solidification and annealing. *J. Alloys Compd.* **2015**, *646*, 503–511. [CrossRef]
49. Krypiakewytsch, P.I.; Zaretschniuk, O.S.; Hladyschewskyj, E.I. Ternäre verbindungen vom NaZn$_{13}$-type. *Z. Anorg. Allg. Chem.* **1968**, *358*, 90–96. [CrossRef]
50. Passamani, E.C.; Larica, C.; Proveti, J.R.; Takeuchi, A.Y.; Gomes, A.M.; Ghivelder, L. Magnetic and magnetocaloric properties of La(Fe,Co)$_{11.4}$SP$_{1.6}$ compounds (SP=Al or Si). *J. Magn. Magn. Mater.* **2007**, *312*, 65–71. [CrossRef]
51. Shao, Y.Y.; Liu, Y.F.; Wang, K.; Zhang, M.X.; Liu, J. Impact of interface structure on functionality in hot-pressed La-Fe-Si/Fe magnetocaloric composites. *Acta Mater.* **2020**, *195*, 163–171. [CrossRef]
52. Zhang, M.X.; Ouyang, Y.; Zhang, Y.F.; Liu, J. LaFe$_{11}$Co$_{0.8}$Si$_{1.2}$/Al magnetocaloric composites prepared by hot pressing. *J. Alloys Compd.* **2020**, *823*, 153846. [CrossRef]
53. Wang, Y.X.; Zhang, H.; Liu, E.K.; Zhong, X.C.; Tao, K.; Wu, M.L.; Xing, C.F.; Xiao, Y.N.; Liu, J.; Long, Y. Outstanding comprehensive performance of La(Fe,Si)$_{13}$H$_y$/In composite with durable service life for magnetic refrigeration. *Adv. Electron. Mater.* **2018**, *4*, 1700636. [CrossRef]
54. Gschneidner, K.A., Jr.; Pecharsky, V.K. Magnetocaloric materials. *Annu. Rev. Mater. Sci.* **2000**, *30*, 387–429. [CrossRef]
55. Fujieda, S.; Fukamichi, K.; Suzuki, S. Microstructure and isothermal magnetic entropy change of La(Fe$_{0.89}$Si$_{0.11}$)$_{13}$ in a single-phase formation process by annealing. *J. Alloys Compd.* **2013**, *566*, 196–200. [CrossRef]
56. Wu, S.M.; Zhong, X.C.; Dong, X.T.; Liu, C.L.; Huang, J.H.; Huang, Y.L.; Yu, H.Y.; Liu, Z.W.; Huang, Y.S.; Ramanujan, R.V. LaFe$_{11.6}$Si$_{1.4}$/Pr$_{40}$Co$_{60}$ magnetocaloric composites for refrigeration near room temperature. *J. Alloys Compd.* **2021**, *873*, 159796. [CrossRef]
57. Yu, P.; Zhang, N.Z.; Cui, Y.T.; Wu, Z.M.; Wen, L.; Zeng, Z.Y.; Xia, L. Achieving better magneto-caloric effect near room temperature in amorphous Gd$_{50}$Co$_{50}$ alloy by minor Zn addition. *J. Non-Crystall. Solids* **2016**, *434*, 36–40. [CrossRef]
58. Dan'kov, S.Y.; Tishin, A.M.; Pecharsky, V.K.; Gschneidner, K.A., Jr. Magnetic phase transitions and the magnetothermal properties of gadolinium. *Phys. Rev. B* **1998**, *57*, 3478–3490. [CrossRef]
59. Zhong, X.C.; Tian, H.C.; Wang, S.S.; Liu, Z.W.; Zheng, Z.G.; Zeng, D.C. Thermal, magnetic and magnetocaloric properties of Fe$_{80-x}$M$_x$B$_{10}$Zr$_9$Cu$_1$ (M = Ni, Ta; x = 0, 3, 5) amorphous alloys. *J. Alloys Compd.* **2015**, *633*, 188–193. [CrossRef]

60. Banerjee, B.K. On a generalised approach to first and second order magnetic transitions. *Phys. Lett.* **1964**, *12*, 16–17. [CrossRef]
61. Hu, F.X. *Magnetic and Magnetic Entropy Changes of Fe-Based La(Fe,M)$_{13}$ Compounds and Ni-Mn-Ga Alloys*; Institute of Physics, Chinese Academy of Sciences: Beijing, China, 2002. (In Chinese)

Recent Advances in Magnetostrictive Tb-Dy-Fe Alloys

Review

Zijing Yang, Jiheng Li *, Zhiguang Zhou, Jiaxin Gong, Xiaoqian Bao and Xuexu Gao

State Key Laboratory for Advanced Metals and Materials, University of Science and Technology Beijing, Beijing 100083, China; g20209193@xs.ustb.edu.cn (Z.Y.); zzgsunrui@126.com (Z.Z.); gongjiaxin416@163.com (J.G.); bxq118@ustb.edu.cn (X.B.); gaox@skl.ustb.edu.cn (X.G.)
* Correspondence: lijh@ustb.edu.cn; Tel.: +86-10-6233-3431

Abstract: As giant magnetostrictive materials with low magnetocrystalline anisotropy, Tb-Dy-Fe alloys are widely used in transducers, actuators and sensors due to the effective conversion between magnetic energy and mechanical energy (or acoustic energy). However, the intrinsic brittleness of intermetallic compounds leads to their poor machinability and makes them prone to fracture, which limits their practical applications. Recently, the addition of a fourth element to Tb-Dy-Fe alloys, such as Ho, Pr, Co, Nb, Cu and Ti, has been studied to improve their magnetostrictive and mechanical properties. This review starts with a brief introduction to the characteristics of Tb-Dy-Fe alloys and then focuses on the research progress in recent years. First, studies on the crystal growth mechanism in directional solidification, process improvement by introducing a strong magnetic field and the effects of substitute elements are described. Then, meaningful progress in mechanical properties, composite materials, the structural origin of magnetostriction based on ferromagnetic MPB theory and sensor applications are summarized. Furthermore, sintered composite materials based on the reconstruction of the grain boundary phase also provide new ideas for the development of magnetostrictive materials with excellent comprehensive properties, including high magnetostriction, high mechanical properties, high corrosion resistance and high resistivity. Finally, future prospects are presented. This review will be helpful for the design of novel magnetostrictive Tb-Dy-Fe alloys, the improvement of magnetostrictive and mechanical properties and the understanding of magnetostriction mechanisms.

Keywords: magnetostriction; Tb-Dy-Fe alloys; directional solidification; mechanical property; Tb-Dy-Fe composites; applications

1. Introduction

The physical effect of the magnetostriction of Tb-Dy-Fe alloys is utilized to realize their application in sensors, transducers and actuators through the conversion of magnetoelastic properties and mechanical energy.

Clark et al. [1,2] discovered that the magnetization and magnetocrystalline anisotropy of composite rare-earth compounds composed of R'Fe$_2$ and R''Fe$_2$ (R' and R'' denote different rare-earth elements) had a superposition effect. In particular, the λ_{111} of pseudobinary Tb$_x$Dy$_{1-x}$Fe$_2$ compounds (0 < x < 1) could reach 1600–2400 × 10^{-6}, and the external magnetic field intensity required to achieve saturation magnetization was only 1.6 × 10^3 kA/m. Due to the large anisotropy of magnetostriction in the Tb-Dy-Fe single crystal, the magnetostrictive strain in the <111> easy axis is the largest. However, <111> is not the easy growth direction of the crystal. It is necessary to develop directional solidification technology to bring the grain orientation closer to the easy magnetization direction <111> [3]. Tb-Dy-Fe alloys in <110> and <112> orientations are usually prepared by directional solidification [4,5].

During the last years, many efforts have been dedicated to enhancing magnetostriction to reduce costs, such as alloying with other elements and improving the preparation process [6–9]. In previous research, the partial substitution of Tb and Dy by Ho was investigated to reduce the magnetocrystalline anisotropy and effectively decrease the hysteresis [10]. Some multicomponent alloys, such as $(Tb_{0.7}Dy_{0.3})_{0.7}Pr_{0.3}(Fe_{1-x}Co_x)_{1.85}$ ($0 \leq x \leq 0.6$) and $Tb_{0.3}Dy_{0.7}(Fe_{1-x}Si_x)_{1.95}$ (x = 0.025), also presented good low-field magnetostriction performance [10–12]. However, intrinsic brittleness and large eddy-current loss at high frequency still limit the application range of Tb-Dy-Fe alloys.

The magnetostriction of Tb-Dy-Fe alloys is related to the magnetocrystalline anisotropy of rare-earth compounds. It is also considered to be derived from the interaction between 4f electrons of rare-earth elements and 3d electrons of transition-metal ions [13]. Since the (Tb, Dy)Fe$_2$ pseudobinary alloy system was proposed, there has been little effective progress in the understanding of its magnetostriction mechanism. To explore the great enhancement of its properties, an in-depth understanding of its physical nature is urgently needed. In recent years, the concept and implication of the morphotropic phase boundary (MPB) have been introduced to the ferromagnetic material system, which provides a new perspective for the research of the magnetostrictive effect of Tb-Dy-Fe alloys and the development of high-performance magnetostrictive materials [14–16]. Furthermore, the emergence and development of a new generation of synchrotron and light sources could more accurately detect the position change of atoms in the crystal, which would be conducive to the study of the magnetostriction mechanism of Tb-Dy-Fe alloys [17,18].

Based on previous works, this review focuses on the research progress of Tb-Dy-Fe alloys in recent years. In Section 2, the study of the crystal growth mechanism in directional solidification and the introduction of a strong magnetic field for process improvement are introduced. The effects of substitute elements are discussed in Section 3, and then the meaningful progress made in recent years in understanding the mechanical properties, composites and ferromagnetic MPB of Tb-Dy-Fe alloys is summarized in Sections 4–6, respectively. Finally, the latest progress in the application of high-sensitivity sensors designed with Tb-Dy-Fe alloys is discussed.

2. Grain Orientation and Properties of Directionally Solidified Tb-Dy-Fe Alloys

In order to achieve the large magnetostriction of Tb-Dy-Fe compounds in a low magnetic field, directional solidification technology needs to be used to orient the grains in the easy magnetization direction as much as possible due to the anisotropy of magnetostriction.

2.1. Grain Growth and Orientation Control during Directional Solidification Process

The growth process of crystals in directional solidification directly affects the final orientation of grains. Therefore, it is necessary to understand the crystal growth mechanism and orientation selection mechanism during this process; for example, the solid–liquid interface morphology and atomic adhesion kinetics should be researched.

The solid–liquid interface morphology plays a key role in single-crystal growth. By controlling the zone-melting length using a modified optical zone-melting method, Kang et al. [19] obtained three forms of solid–liquid interface morphologies, namely, convex, flat and concave interfaces, as shown in Figure 1. They successfully prepared an <110> axial oriented Tb-Dy-Fe twinned single crystal without radial composition segregation. Although the convex interface was conducive to single-crystal growth, it would produce radial component segregation; that is, the shape of the solid–liquid interface could affect the radial component distribution. By establishing their theoretical models, the effects of the temperature gradient, growth rate and zone-melting length on radial component segregation were qualitatively described.

Generally, solidification parameters, such as the temperature gradient and solidification rate, influence the evolution of texture during the directional solidification process. By comparing the texture at different distances from the onset of solidification, Palit et al. [20] found that the transition in the preferred growth direction from <110> to <112> occurred through intermediate <123> texture components. Furthermore, plane-front solidification morphology and irregular peritectic coupled growth were observed in a wide range of solidification rates (5–80 cm/h), while the preferred direction changed from <311> to

<110> to <112> at 5–100 cm/h solidification rates [21]. The {111} planes in Figure 2a were composed of two different types of atomic layers, in which one layer was all Fe, which increased the obstacle of atomic arrangement in the growth process. At the same time, the {311} planes could be attached to the two {111} planes, as shown in Figure 2b. Therefore, although {111} planes had higher atomic bulk density, the preferred orientation at low solidification rates (5–30 cm/h) was <311>. In addition, the highest magnetostriction was achieved in the sample with a solidification rate of 100 cm/h due to the <112> preferred orientation, as shown in Figure 2c,d.

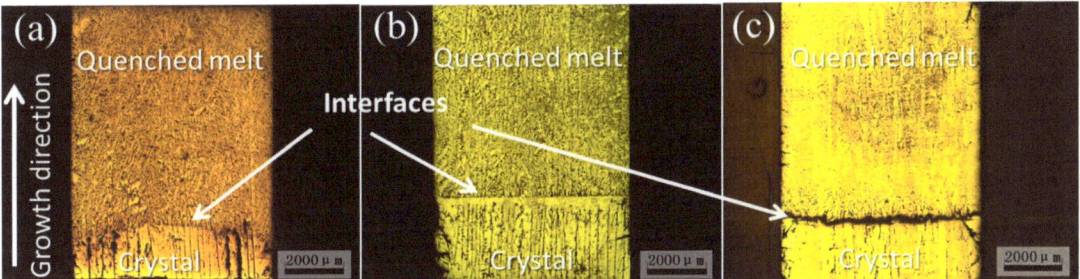

Figure 1. Three forms of solid–liquid interface morphologies under the condition of V = 15 mm/h with different half zone-melting lengths: (**a**) P = 60; $L_{1/2}$ = 7 mm (convex); (**b**) P = 70; $L_{1/2}$ = 10 mm (flat); (**c**) P = 80; $L_{1/2}$ = 15 mm (concave) [19] (here, P is heating power) (Reprinted with permission from Ref. [19]. Copyright 2015 Elsevier).

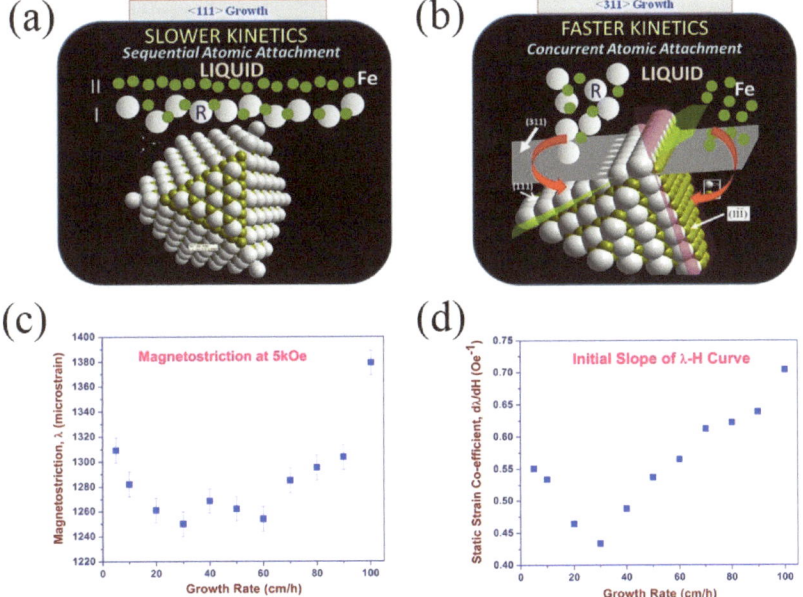

Figure 2. Comparison of atomic attachment kinetics at (**a**) (111) and (**b**) (311) interfaces; (**c**) magnetostriction (λ) of directionally solidified samples measured at an applied field of 5 kOe and (**d**) the plot of slope (dλ/dH) of the initial λ-H plot, plotted as a function of growth rate [21] (Adapted with permission from Ref. [21]. 2016 Springer Nature).

The growth twins of Tb-Dy-Fe alloys changed the crystal orientation, which was related to the crystal orientation of mirror symmetry [22]. Previous studies have shown that different solidification rates correspond to different solidification morphologies; that is, with the increase in the solidification rate, the preferred axial orientation changed from <101> to <112> and was then reoriented to <101> [23]. Recently, 3D spatial extension and transformation of the whole grain were asserted to be the key to interpreting the transformation of the preferred axial orientation. As a polyhedral material with a face-centered cubic structure, Tb-Dy-Fe alloy dendrites grow in the form of twin-related lamellae. In this case, when the crystal grows in a cellular form, the initial dendritic arms of the <110> axially oriented crystal have two extension directions, which will occupy more space and obtain preferential growth (Figure 3). Therefore, the transformation of the preferred axial orientation was explained by the different space-occupying capacities caused by the different morphological configurations for <101> and <112> axially oriented grains [24].

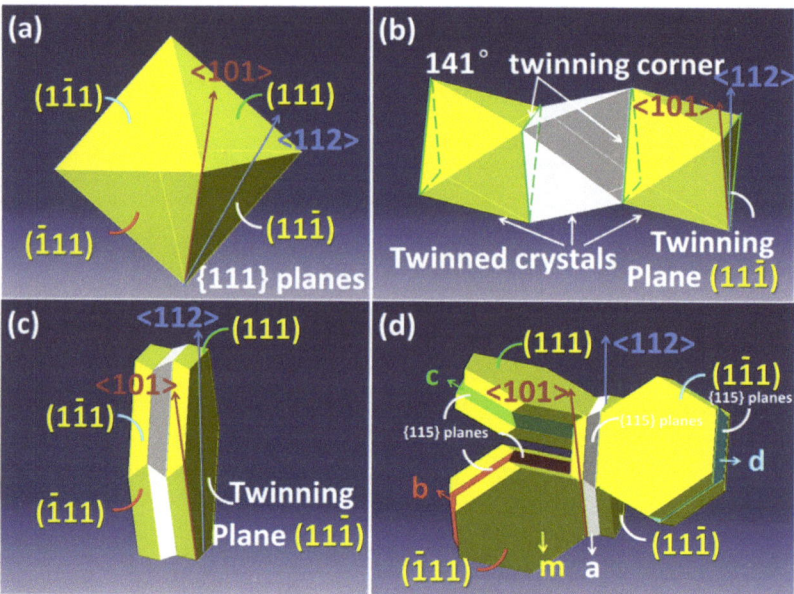

Figure 3. Octahedral configuration: (**a**) a faceted equiaxed grain m bounded by eight {111} planes; (**b**) occurrence of double parallel twin boundaries on the (11$\bar{1}$) plane of the crystal; (**c**) twinned lamellas grow on the (11$\bar{1}$) plane; (**d**) twinned lamellas grow on four different {111} planes [24]. Reprinted with permission from ref. [24]. Copyright 2018 Elsevier.

2.2. <111>-Oriented Tb-Dy-Fe Alloys Prepared by Directional Solidification in Magnetic Fields

Aiming to prepare Tb-Dy-Fe alloys with preferred orientation along <111> or close to <111>, researchers introduced a strong magnetic field to induce the crystal orientation in a specific direction during the directional solidification process. According to the relevant theory of crystal orientation induced by a magnetic field [25], if there is sufficient action time and rotation space, grains of materials with magnetocrystalline anisotropy will rotate and be oriented under the action of a strong magnetic field through Lorentz force, magnetic force and the magnetic moment of materials.

A high magnetic field in the horizontal direction was directly applied to the directional solidification process [26]. Liu et al. [27–31] obtained a higher <111> orientation and improved magnetostrictive properties by applying a constant magnetic field of 4.4 T during the solidification process of $Tb_{0.27}Dy_{0.73}Fe_{1.95}$ alloy, as shown in Figure 4. In addition, the best contrast of the domain image and the widest magnetic domain were obtained in the sample prepared under a 4.4 T magnetic field [27]. The required magnetic field in the range of 4–10 T to achieve <111> orientation increased with the increase in the cooling rate [28].

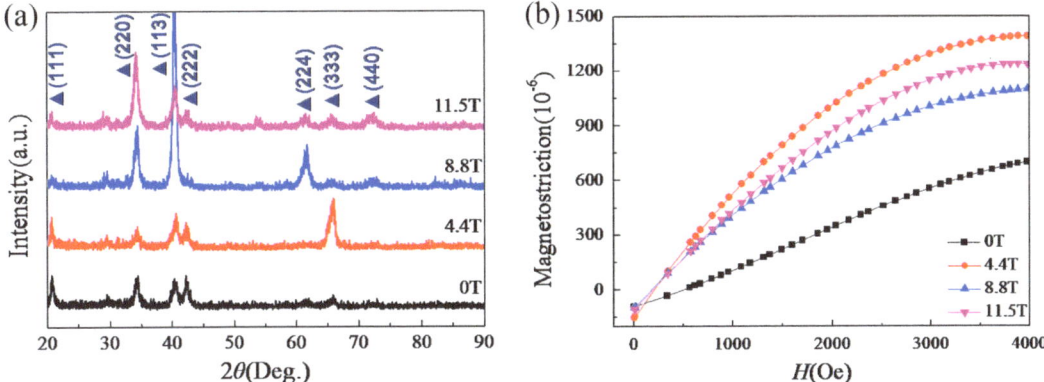

Figure 4. (**a**) XRD patterns in the plane perpendicular to the magnetic field direction and (**b**) magnetostriction for Tb$_{0.27}$Dy$_{0.73}$Fe$_{1.95}$ alloys solidified in various high magnetic fields [27]. Reprinted with permission from Ref. [27]. Copyright 2016 Elsevier.

Recently, researchers obtained both orientation and alignment along <111> by multiple magnetic field effects of the liquid phase and solid phase, as shown in Figure 5 [29,30]. In addition, magnetostrictive and mechanical properties were increased by adjusting the content, morphology and distribution of the (Tb, Dy)Fe$_3$ phase and WSP by coupling directional solidification with a high magnetic field [31].

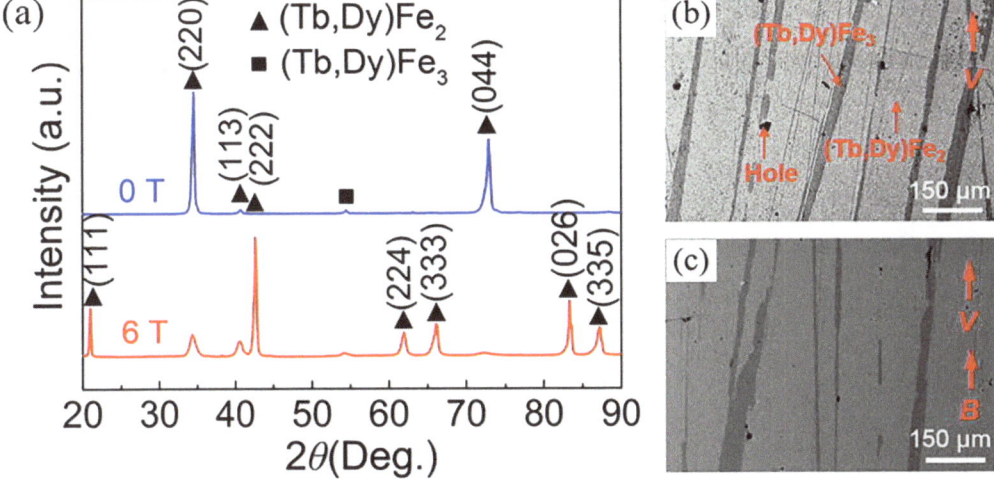

Figure 5. (**a**) XRD patterns of Tb$_{0.27}$Dy$_{0.73}$Fe$_{1.95}$ alloys on the transverse section; (**b**,**c**) SEM images of the alloy structures grown with (**b**) 0T and (**c**) 6 T magnetic field [29]. Reprinted with permission from ref. [29]. Copyright 2020 AIP Publishing.

Furthermore, when a gradient magnetic field strongly dependent on the cooling rate was applied during cooling and solidification, magnetic gradient Tb$_{0.27}$Dy$_{0.73}$Fe$_{1.95}$ alloy with gradient magnetostriction and saturation magnetization was obtained, as shown in Figure 6, which was attributed to the increase in the gradient of the orientation degree [32–34].

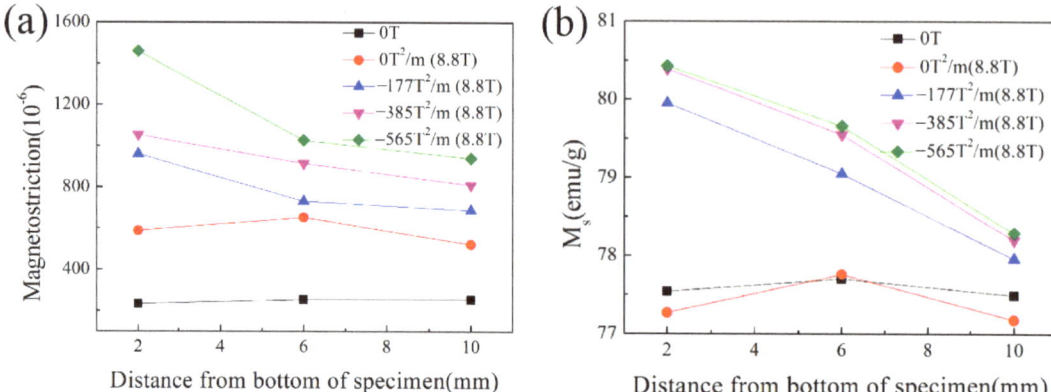

Figure 6. (a) The maximum magnetostriction of alloys solidified in various high magnetic fields at 4000 Oe and (b) saturation magnetization through the depths of alloys solidified in various high magnetic fields [32]. Reprinted with permission from Ref. [32]. Copyright 2016 World Scientific Publishing Company.

3. Effects of Substitute Elements on Magnetostriction of Tb-Dy-Fe Alloys

Cubic RFe$_2$ compounds usually have large magnetostriction at room temperature, but their magnetocrystalline anisotropy is also large. In this case, a strong external magnetic field is frequently required to achieve high performance in practical applications, which hinders the application of magnetostrictive RFe$_2$ materials. Initially, aiming to reduce magnetocrystalline anisotropy as well as maintain large magnetostriction, pseudobinary RR'Fe$_2$ compounds with an anisotropy compensation function were developed, that is, by alloying RFe$_2$ compounds with the same magnetostriction symbol but opposite anisotropy symbol. Furthermore, with a Laves phase structure, the magnetic properties of this alloy can be easily changed by various substitutions in rare-earth and 3d transition-metal sublattices.

3.1. Alloy System Containing Other Rare-Earth Elements

The introduction of the third rare-earth element can provide additional degrees of freedom for the (Tb,Dy)Fe$_2$ system, minimizing the first-order anisotropy constant (K_1) and the second-order anisotropy constant (K_2), which is considered as an approach to improve the magnetostriction of Tb-Dy-Fe alloys.

3.1.1. Pr

PrFe$_2$ has a very high magnetostrictive coefficient (about 5600 ppm at 0 k), and the high Pr content is also conducive to anisotropic compensation. However, Tb$_{1-x}$Pr$_x$Fe$_2$ is a noncubic phase when the Pr content exceeds 20% due to the large Pr^{3+} radius, while high Pr content is conducive to anisotropic compensation [13]. During the last years, some studies have confirmed that (Tb, Pr, Dy) Fe$_2$ series compounds are an anisotropic compensation system [35–37].

One way to improve the performance of the (Tb, Pr, Dy)Fe$_2$ system is to replace Fe by adding Co, B and other elements. For instance, (Tb$_{0.7}$Dy$_{0.3}$)$_{0.7}$Pr$_{0.3}$(Fe$_{1-x}$Co$_x$)$_{1.85}$ ($0 \leq x \leq 0.6$) was composed of a MgCu$_2$-type C15 cubic Laves phase, with a small amount of a PuNi$_3$-type phase and rare-earth-rich phase [11]. The second phase of Dy$_{1-x}$(Tb$_{0.2}$Pr$_{0.8}$)$_x$Fe$_{1.93}$ ($0 \leq x \leq 0.5$) obtained by atmospheric pressure annealing appeared when x exceeded 0.3 [37]. Shi et al. successively synthesized Pr$_x$Tb$_{1-x}$Fe$_{1.9}$ ($0 \leq x \leq 1$) [38], Pr(Fe$_{1-x}$Co$_x$)$_{1.9}$ ($0 \leq x \leq 0.5$) [39], Dy$_{1-x}$Pr$_x$Fe$_{1.9}$ ($0 \leq x \leq 1$) [40] and Pr$_{1-x}$Dy$_x$(Fe$_{0.8}$Co$_{0.2}$)$_{1.93}$ (x = 0.00, 0.05, 0.10, 0.20 and 0.30) [36] single cubic Laves compounds by high-pressure annealing. The magnetostriction of Pr$_{0.95}$Dy$_{0.05}$(Fe$_{0.8}$Co$_{0.2}$)$_{1.93}$ alloy at 3kOe was 648 ppm, which is twice that of Tb$_{0.2}$Dy$_{0.58}$Pr$_{0.22}$(Fe$_{0.9}$B$_{0.1}$)$_{1.93}$ (about 300 ppm) [35].

The 440$_C$ XRD profiles and magnetostriction of (Tb$_{0.2}$Pr$_{0.8}$)$_x$Dy$_{1-x}$Fe$_{1.93}$ (x = 0.00, 0.05, 0.10, 0.20 and 0.30) single Laves phase compounds synthesized by high-pressure annealing are shown in Figure 7 [41]. Based on the consideration of anisotropy compensation and thermodynamic energy flattening, a rare-earth sublattice was designed, and the ternary composition phase diagram and the minimum anisotropy composition are shown in Figure 8 [42].

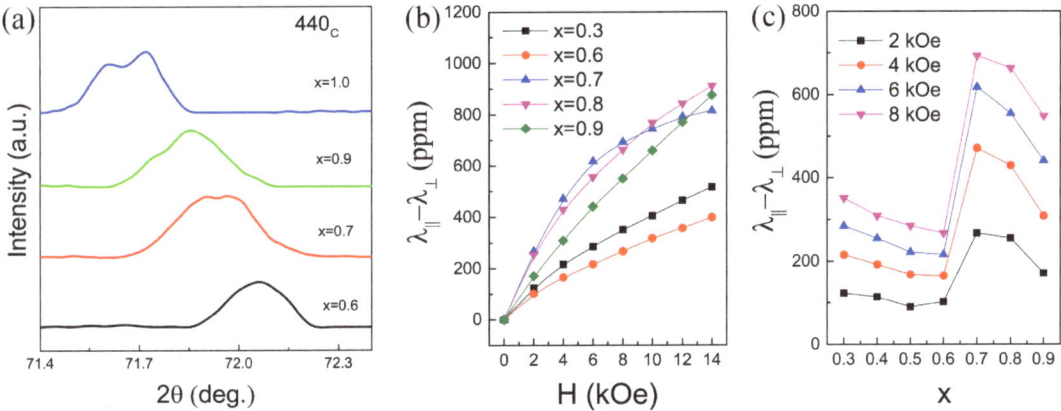

Figure 7. (a) Typical step-scanned 440$_C$ XRD profiles of the (Tb$_{0.2}$Pr$_{0.8}$)$_x$Dy$_{1-x}$Fe$_{1.93}$ Laves phase, and room temperature magnetostriction $\lambda_\parallel - \lambda_\perp$ as a function of (b) the applied field and (c) the composition [41]. Reprinted with permission from Ref. [41]. Copyright 2019 Elsevier.

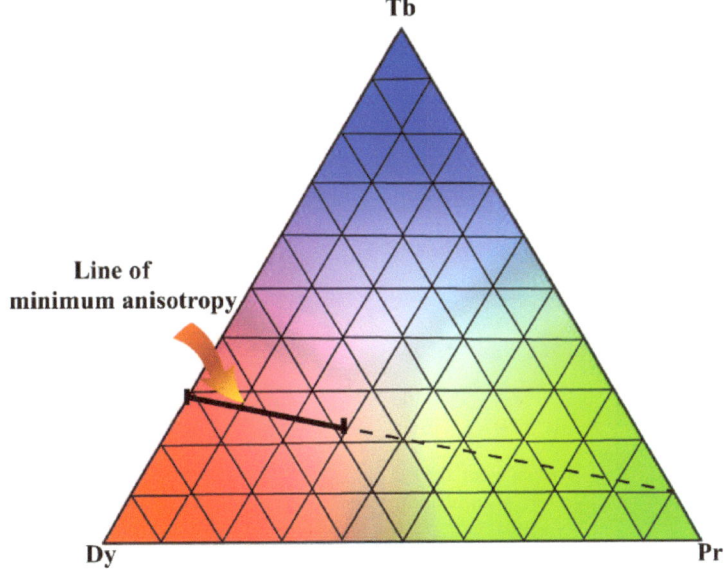

Figure 8. Designed ternary composition phase diagram for the Tb-Dy-Pr system: the location between two bias diagonals of the rare-earth composition is indicated along the expected minimum anisotropy line [42]. Reprinted with permission from Ref. [42]. Copyright 2020 Springer Nature.

3.1.2. Nd

Similar to Pr, many studies [43–52] have shown that the appropriate replacement of heavy rare-earth Tb and Dy with light rare-earth Nd is an effective approach to decrease magnetocrystalline anisotropy and improve magnetostriction. For example, the $Tb_{0.4}Dy_{0.5}Nd_{0.1}Fe_{1.93}$ compound possessed the large magnetostriction value of 1300 ppm at 5 kOe [47]. Pan et al. [45,48] obtained good magnetoelasticity in $Tb_{0.3}Dy_{0.6}Nd_{0.1}(Fe_{0.8}Co_{0.2})_{1.93}$ by improving the annealing process.

$Tb_{0.2}Nd_{0.8}(Fe_{0.8}Co_{0.2})_{1.9}$ ribbons with high Nd content were prepared by melt spinning and low-temperature annealing, which provided another effective method for the synthesis of a C15 cubic Laves phase structure with high Nd content [49]. The results indicated that a higher solidification rate is conducive to the elimination of the (Tb, Nd)Fe_3 phase. A single cubic Laves phase with a <111> easy magnetization direction at room temperature was obtained at 45 m/s runner speed and 773 K annealing temperature [50]. Subsequently, a series of $Tb_{0.2}Nd_{0.8}(Fe_{1-x}Co_x)_{1.9}$ ($0 \leq x \leq 0.4$) compounds were fabricated by rapid melt quenching in order to study the effects of Co substitution for Fe. It was found that the lattice parameter decreased with increasing x, and the ($\lambda_\parallel - \lambda_\perp$) of ribbons with x = 0.1–0.2 at 10 kOe was 306 ppm and 321 ppm, respectively [51]. That is, an appropriate amount of Co instead of Fe could promote the formation of a single Laves phase and improve the magnetostrictive properties of $Tb_{0.2}Nd_{0.8}(Fe_{1-x}Co_x)_{1.9}$ ribbons.

Recently, the spin configuration phase diagram (Figure 9a) of $Tb_{0.27}Dy_{0.73-x}Nd_xFe_2$ ($0 \leq x \leq 0.4$) was designed based on the experimental results of magnetization with varying temperature, magnetic susceptibility and XRD analysis [52]. In this case, it was found that the Curie temperature, spin reorientation temperature, saturation magnetization and magnetostriction (Figure 9b) of the $Tb_{0.27}Dy_{0.73-x}Nd_xFe_2$ compound decreased with the increase in Nd concentration. Moreover, a value of λ_{111} = 1700 ppm in the $Tb_{0.27}Dy_{0.63}Nd_{0.1}Fe_2$ Laves phase compound was obtained by high-pressure annealing.

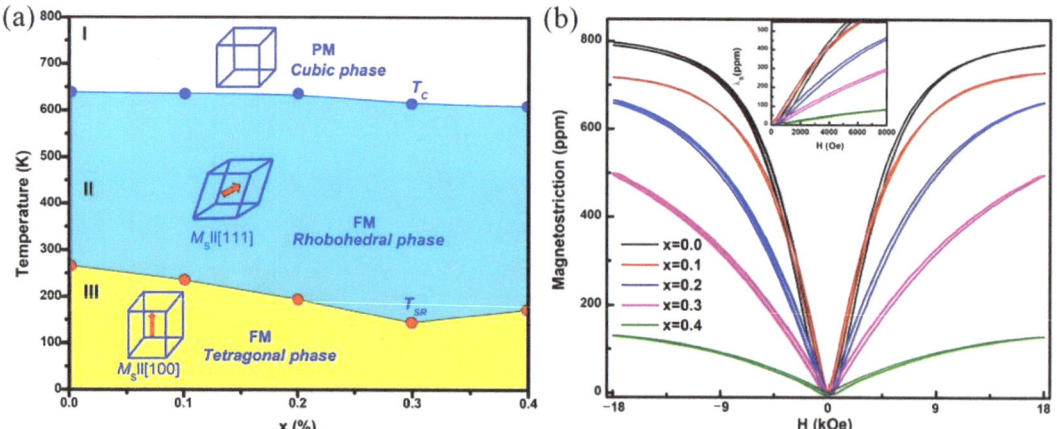

Figure 9. For $Tb_{0.27}Dy_{0.73-x}Nd_xFe_2$: (**a**) spin configuration phase diagram accompanied by the cubic, rhombohedral and tetragonal crystal symmetries and (**b**) room temperature magnetostriction (λ_S) (x = 0.0, 0.1.0.2, 0.3, 0.4) [52]. Reprinted with permission from ref. [52]. Copyright 2020 Elsevier.

3.1.3. Ho

Substituting a small amount of Ho for Tb and Dy in $Tb_{0.3}Dy_{0.7}Fe_2$ can significantly decrease anisotropy and hysteresis loss [53]. Moreover, the addition of Ho also narrowed the temperature range between the liquidus temperature and the peritectic temperature, which was beneficial for reducing the pre-peritectic of the (Tb, Dy, Ho) Fe_3 phase. $Tb_{0.26}Dy_{0.49}Ho_{0.25}Fe_{1.9}$ had both large magnetostriction and small hysteresis under a low magnetic field. Furthermore, it has also been found that magnetic annealing effectively increases magnetostriction [54]. Figure 10 shows the anisotropic compensation effects of Ho and Pr in $Tb_{0.1}Ho_{0.9-x}Pr_x(Fe_{0.8}Co_{0.2})_{1.93}$ alloys and the effects of their components on the phase structure and magnetostriction [55].

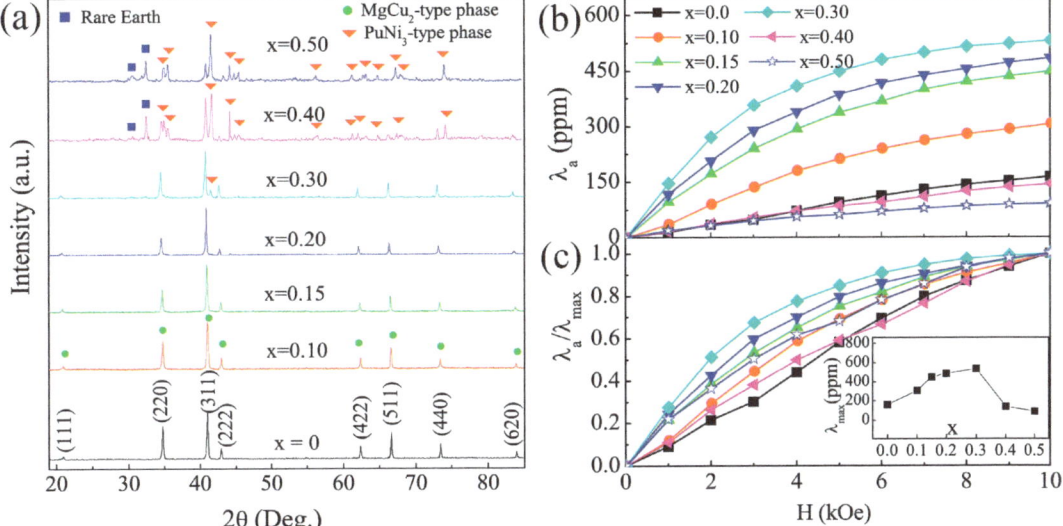

Figure 10. (**a**) XRD patterns and magnetic-field dependence of (**b**) magnetostriction and (**c**) normalized magnetostriction for $Tb_{0.1}Ho_{0.9-x}Pr_x(Fe_{0.8}Co_{0.2})_{1.93}$ alloys [55]. Reprinted with permission from Ref. [55]. Copyright 2016 Elsevier.

Moreover, the addition of Ho reduced the saturated magnetic field (220 kA/m) and dynamic magnetic loss of Tb-Dy-Fe fiber composites [56], and the maximum magnetostriction of the Tb-Dy-Ho-Fe/epoxy composite was 695 ppm when the Ho content x = 0.31 [57].

3.2. Alloy System Containing Other Elements

In a system containing light rare-earth elements, an appropriate amount of Co instead of Fe promoted and stabilized the formation of the cubic Laves phase [51,58,59]. In addition, 20 at% Co instead of Fe was found to increase the Curie temperature and saturation magnetization and improve the magnetostriction of $Tb_{0.2}Nd_{0.8}Fe_{1.9}$ ribbons [51]. Co substitution for Fe was able to extend the operating temperature range for $Tb_{0.36}Dy_{0.64}Fe_2$ by increasing the Curie temperature (T_c) or decreasing the spin reorientation temperature (T_r) [60,61]. Recently, Yang et al. found a new "Griffiths-like transition" in $Tb_{0.3}Dy_{0.7}(Co_{1-x}Fe_x)_2$ when x < 0.8 and suggested that its disappearance was due to the interaction between Fe and Co [62]. In addition, enhanced magnetostriction (818 ppm) and high T_c (707 K) were obtained at x = 0.8. Theoretically, the doping of transition metals modulated the exchange action between 3d-3d and 3d-4f atoms, and it is expected to improve the elastic energy and magnetostatic energy.

Based on the different effects of the addition of elements on properties, Wang et al. [59] divided different elements into two types: those that readily formed phases with rare-earth elements and those that readily formed phases with the enthalpy of mixing between atomic pairs. After adding Nb, Ti and V elements, the second phases $NbFe_2$, Fe_2Ti and FeV, respectively, were dispersed in the Tb-Dy-Fe matrix alloy, as shown in Figure 11, which can inhibit the formation of the harmful RFe_3 phase.

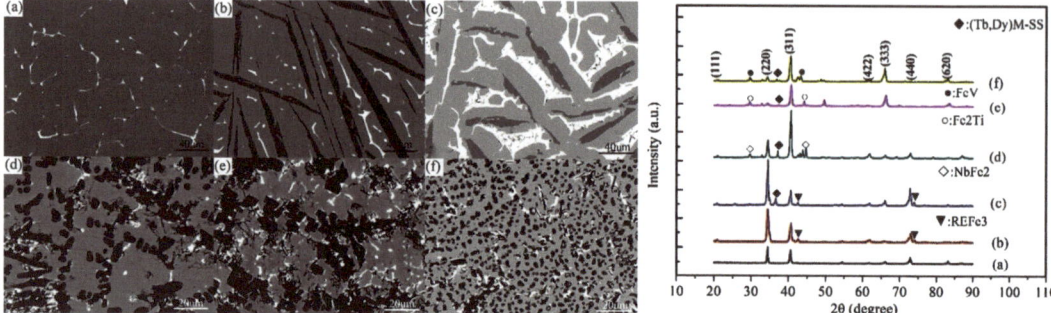

Figure 11. BSEM images and X-ray diffraction patterns of the alloys (**a**) $Tb_{0.3}Dy_{0.7}Fe_2$; (**b**) $(Tb_{0.3}Dy_{0.7})_{0.3}Fe_{0.6}Co_{0.1}$; (**c**) $(Tb_{0.3}Dy_{0.7})_{0.3}Fe_{0.6}Cu_{0.1}$; (**d**) $(Tb_{0.3}Dy_{0.7})_{0.3}Fe_{0.6}Nb_{0.1}$; (**e**) $(Tb_{0.3}Dy_{0.7})_{0.3}Fe_{0.6}Ti_{0.1}$; (**f**) $(Tb_{0.3}Dy_{0.7})_{0.3}Fe_{0.6}V_{0.1}$ [59]. Reprinted with permission from ref. [59]. Copyright 2018 Elsevier.

4. Mechanical Properties of Tb-Dy-Fe Alloys

The $MgCu_2$-type RFe_2 phase in the alloy provides a large magnetostrictive coefficient, but at the same time, the Laves phase is brittle and fractures easily at room temperature because of its topological close packing structure and lack of available slip system. There are also a large number of parallel acicular Widmanstatten precipitates (WSPs) in the Laves phase, that is, tiny $(Tb, Dy)Fe_3$ phases [63,64]. The complex staggered distribution of WSPs and the lamellar distribution of $(Tb, Dy)Fe_3$ phases in the matrix will adversely affect the mechanical properties.

As the Fe content decreases from a stoichiometric ratio of 2.0, the strength increases significantly, because the ductile rare-earth phase serving as the skeleton network delays crack propagation in the brittle matrix [65]. Heat treatment [6,66] can also improve the mechanical properties by controlling the dispersion and uniform distribution of the spherical rare-earth-rich phase.

In studies on alloying, the $NbFe_2$ phase [67] and $(Tb, Dy)Cu$ phase [68] were found to have the ability to prevent crack propagation, which was beneficial for the improvement of the mechanical properties of the Tb-Dy-Fe alloys. It was suggested that the existence of a soft phase in the alloy made the material exhibit inelastic strain under tensile or shear load. For instance, the soft $(Tb, Dy)Cu$ phase played a key role in stopping or changing the direction of cracks, as shown in Figure 12. The addition of Cu increased the fracture toughness by 2–3 times, and the alloy with 1 at% Cu showed the best fracture toughness of 3.47 $MPa \cdot m^{1/2}$. With the addition of Nb, the fracture toughness was 1.5–5 times higher than that of Nb-free alloy.

Inspired by the ductility $(Tb, Dy)Cu$ phase, the low-melting-point Dy-Cu alloy was introduced to the grain boundary phase of directionally solidified $Tb_{0.3}Dy_{0.7}Fe_{1.95}$ alloy by grain boundary diffusion [69]. The results revealed that the magnetostrictive properties were maintained at 1021–1448 ppm, and the optimum bending strength was increased by nearly 2.6 times (Figure 13). The large magnetostriction was attributed to the matrix Laves phase structure and stable preferred orientation during the grain boundary diffusion process. In addition, the grain boundary phase was mainly composed of the ductile $(Dy,Tb)Cu$ phase, which can retard crack propagation and improve the mechanical strength of $Tb_{0.3}Dy_{0.7}Fe_{1.95}$ alloy.

The content, morphology and distribution of the $(Tb, Dy)Fe_3$ phase and WSP in $Tb_{0.3}Dy_{0.7}Fe_{1.95}$ alloy can be controlled by the effect of a magnetic field on grain orientation and element diffusion during the solidification process [31]. When the angle between the $(Tb, Dy)Fe_3$ phase and the grain growth direction was the smallest and the WSP content was low, the alloy had better mechanical properties. The application of a 6T magnetic field during the solidification process improved the mechanical properties of the alloy at the same growth rate, as shown in Figure 14.

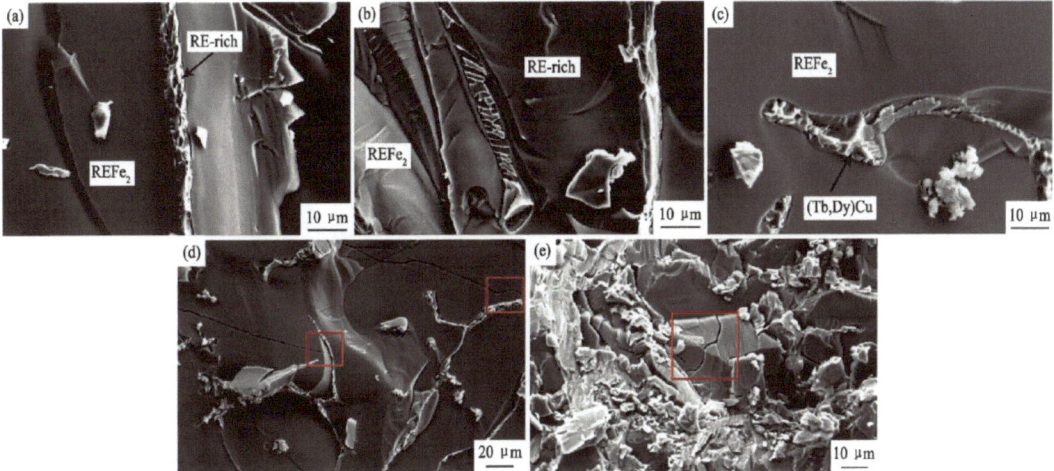

Figure 12. SEM images of crack propagation and soft phase deformation during the compressive fracture process in alloys. (**a**,**b**) (Tb$_{0.3}$Dy$_{0.7}$)$_{0.37}$Fe$_{0.63}$; (**c**,**d**) (Tb$_{0.3}$Dy$_{0.7}$)$_{0.37}$Fe$_{0.62}$Cu$_{0.01}$; (**e**) (Tb$_{0.3}$Dy$_{0.7}$)$_{0.37}$Fe$_{0.53}$Cu$_{0.1}$ [68]. Reprinted with permission from ref. [68]. Copyright 2019 Elsevier.

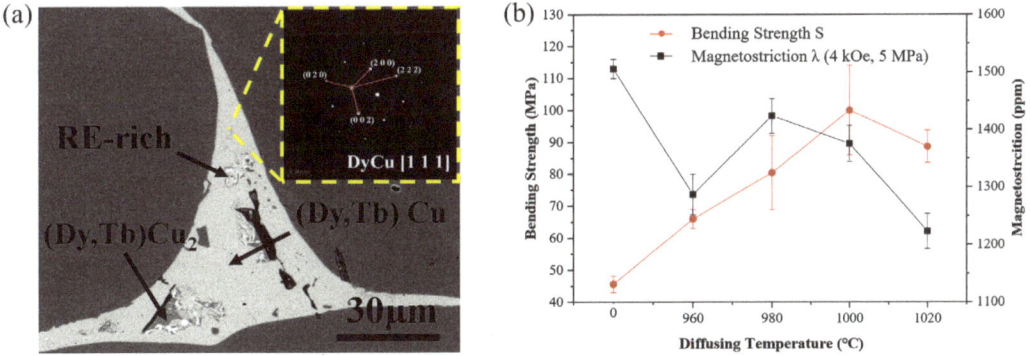

Figure 13. (**a**) SEM images and (**b**) bending strength and magnetostrictive properties of the Tb$_{0.3}$Dy$_{0.7}$Fe$_{1.95}$ alloy diffused by DyCu$_2$ alloy at 980 °C for 3 h, followed by quenching to room temperature; the inset in (**a**) is SAED pattern of DyCu along the [111] zone axis [69]. Reprinted with permission from ref. [69]. Copyright 2020 Elsevier.

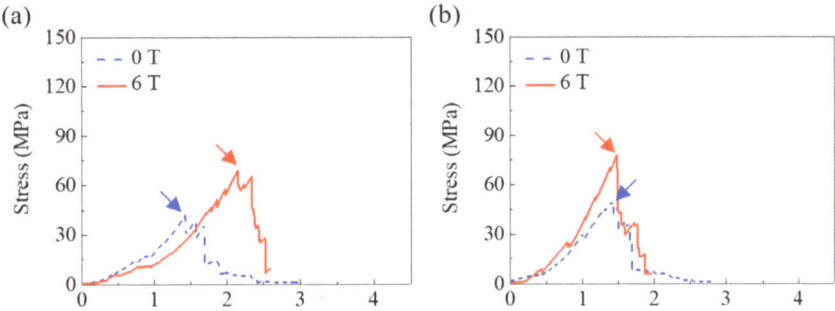

Figure 14. *Cont.*

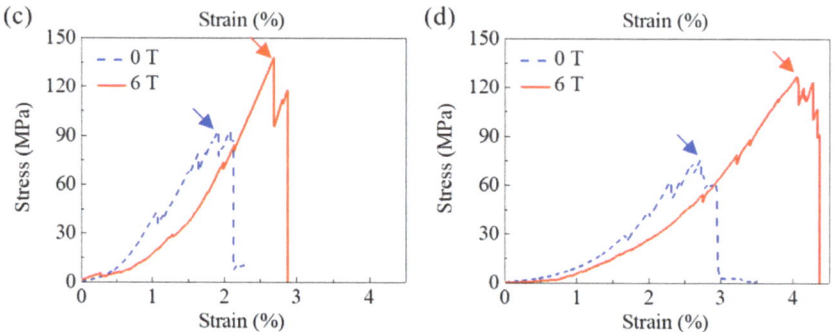

Figure 14. Stress–strain curves of $Tb_{0.27}Dy_{0.73}Fe_{1.95}$ alloys that were solidified directionally at various growth velocities without and with a 6 T magnetic field: (**a**) 25 μm/s; (**b**) 50 μm/s; (**c**) 100 μm/s; (**d**) 200 μm/s [31]. Reprinted with permission from ref. [31]. Copyright 2020 Elsevier.

5. Structural Origin and Magnetic Morphotropic Phase Boundary (MPB) of Tb-Dy-Fe Alloys

To explore the more efficient Tb-Dy-Fe giant magnetostrictive material system, it is necessary to understand the basic principle of magnetostriction, rather than only having a phenomenological understanding. During the last years, with the development of high-resolution synchrotron radiation X-ray diffraction technology, it has been increasingly recognized that the ferromagnetic phase transition process is accompanied by a change in structural symmetry. Significantly, in-depth research of the magnetic morphotropic phase boundary (MPB) has been carried out. In previous research, the synchrotron radiation X-ray diffraction data of $Tb_{0.3}Dy_{0.7}Fe_2$ at 300 K revealed that the {440} and {222} reflections were obviously split, reflecting rhombic symmetry and lattice stretching along the [111] direction. Structurally, this was the process of spin redirection from <001> to <111>, that is, the transition from the T phase with small lattice distortion along the <001> direction to the R phase with large lattice elongation along the <111> direction [70].

The phase diagram of $Tb_{1-x}Dy_xFe_2$ (Figure 15) was obtained through magnetometry and synchrotron XRD experiments [71–73]. It indicates that ferromagnetic MPB is composed of two crystal structures of the parent compounds $TbFe_2$ and $DyFe_2$, with a broadening MPB width at higher temperatures. Furthermore, a simulation based on the energy model demonstrated that the exchange energy narrowed the MPB region by affecting the magnetic phase transition process. This could also be used to explain the above abnormal phenomenon [72]. The exchange interaction was weakened with the increase in temperature, which corresponded to the broadening of the MPB region. In particular, the best point of magnetomechanical application was not centered on MPB but on one side of the rhombohedron. In addition, this local rhombohedral symmetry was further proved by high-resolution transmission electron microscopy. The local nanodomains of ferromagnetic rhombohedral and tetragonal phases coexist in $Tb_{0.3}Dy_{0.7}Fe_2$, as shown in Figure 16 [74]. This is similar to the hierarchical nanodomain structure in ferroelectric materials.

The local stress environment generated by these randomly distributed tetragonal nanodomains in the rhombohedral matrix affected the interaction between Fe_1 and Fe_2 atoms and caused anomalies in the lattice, as shown in Figure 17 [18]. The weak Fe_1-Fe_2 bond was sensitive to the environment, which played an important role in the lattice characteristics of the rhombohedral phase. As a result, the lattice became more orderly with the increase in Dy and more stable Dy-rich phase, which was confirmed by X-ray absorption spectroscopy (XAS) techniques (Figure 18) [18].

These studies also enriched the understanding of the magnetic structure and the origin of large magnetostriction of Tb-Dy-Fe alloys. In detail, the maximum magnetostriction was related to the transition from the T phase with a smaller lattice along the <001> direction to the R phase with larger lattice elongation along the <111> direction.

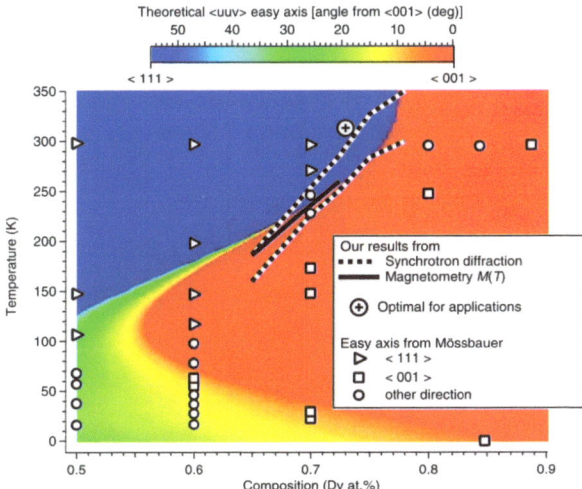

Figure 15. Phase diagram of $Tb_{1−x}Dy_xFe_2$. The background shading shows the magnetic easy axis direction calculated using anisotropy parameters from crystal field theory (see text) [71]. Overlayed is the morphotropic phase boundary determined from our synchrotron XRD (dotted lines) and magnetometry (solid line) measurements, as well as the easy axes reported previously on the basis of Mössbauer spectroscopy (open symbols) [73]. A cross in a circle indicates the optimal temperature (40 °C) for magnetomechanical device applications, as determined for $Tb_{1−x}Dy_xFe_2$ x = 0.73 [2].

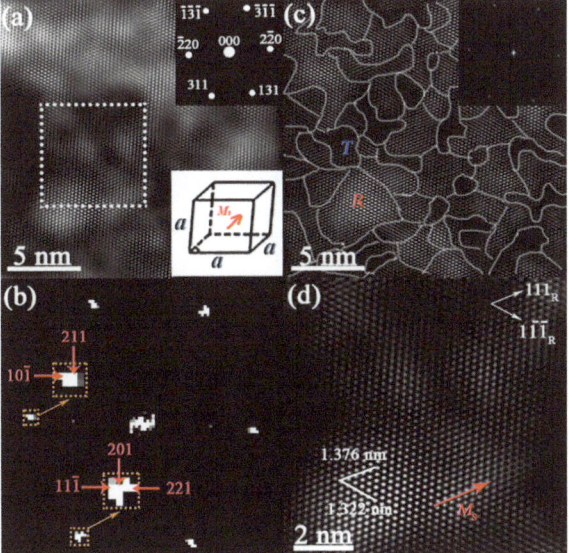

Figure 16. Domains of $Tb_{0.3}Dy_{0.7}Fe_2$ at 293 K revealed by HRTEM [74]. (**a**) HRTEM image taken along the [1$\bar{1}$4]C incident direction. Upper inset is the SAED pattern, and bottom inset is the schematic unit cell with spontaneous magnetization direction. (**b**) FFT of the white rectangle in (**a**) shows splitting reflection spots due to rhombohedral lattice distortion. (**c**) IFFT image by using the {311}C/{111}R reflections in the bottom inset of (**a**), corresponding to the same area in (**a**). The inset is the corresponding FFT spectrum. (**d**) IFFT image by using the {311}C/{111}R reflections from the FFT in (**b**). Adapted with permission from ref. [74]. Copyright 2014 Elsevier.

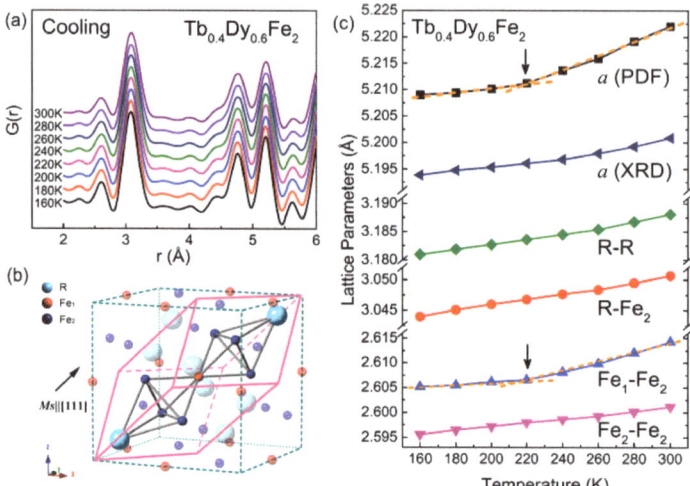

Figure 17. (a) Temperature-dependent pair distribution function (PDF) for $Tb_{0.4}Dy_{0.6}Fe_2$; (b) the rhombohedral cell (pink solid) and cubic cell (teal dashed) for $Tb_{1-x}Dy_xFe_2$; (c) temperature-dependent lattice constants and bond lengths for $Tb_{0.4}Dy_{0.6}Fe_2$ [18]. Reprinted with permission from ref. [18]. Copyright 2020 AIP Publishing.

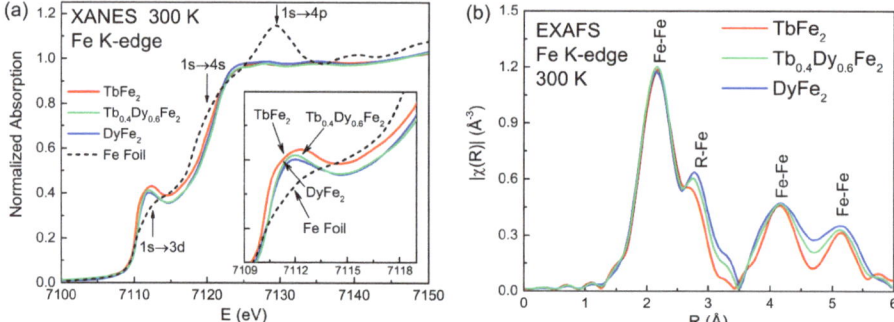

Figure 18. (a) Room temperature Fe K-edge XANES spectra for Fe foil and $Tb_{1-x}Dy_xFe_2$ samples. (b) Room temperature Fe K-edge EXAFS spectra for $Tb_{1-x}Dy_xFe_2$ [18]. Reprinted with permission from ref. [18]. Copyright 2020 AIP Publishing.

It was proposed that the change in the magnetization direction under the action of an external magnetic field could be realized by field-preferred domain growth, which could contribute to explaining large magnetostriction in the low field [75]. The comparison of the diffraction peak intensity of $TbFe_2$ and $Tb_{0.4}Dy_{0.6}Fe_2$ compounds is shown in Figure 19. This magnetostriction of $Tb_{0.4}Dy_{0.6}Fe_2$ was considered to be related to the reduction in rhombohedral distortion caused by replacing Tb by Dy, resulting in field-induced domain conversion, which was more sensitive to the external field [75].

The domain structure and transition near ferromagnetic MPB are also of great significance to understanding the large magnetostriction for Tb-Dy-Fe alloys. The micromechanism of domain strain behavior near ferromagnetic MPB was intuitively illustrated by the phase-field method, combining micromagnetic and micro-elastic theory. This large magnetostrictive strain was considered to be due to the low-energy rotation path of the local magnetization vector in the phase coexistence region. In particular, the tetragonal phase as the intermediate phase provided a low-energy rotation channel for the diamond phase domain from other directions to the external field direction [76,77]. Similar to ferroelectrics, it was believed that the flattening of thermodynamic energy should lead to the sensitive response of the ferromagnetic phase. Through the introduction of $Tb_{0.1}Pr_{0.9}$ to Co-doped $Tb_{0.27}Dy_{0.73}Fe_2$ alloy, Hu et al. explored the feasibility of this assumption through phase-field simulation [42]. They provided a feasible strategy for the design of an ultrasensitive magnetostrictive response with minor metastable orthorhombic phases as bridging domains, as shown in Figure 20.

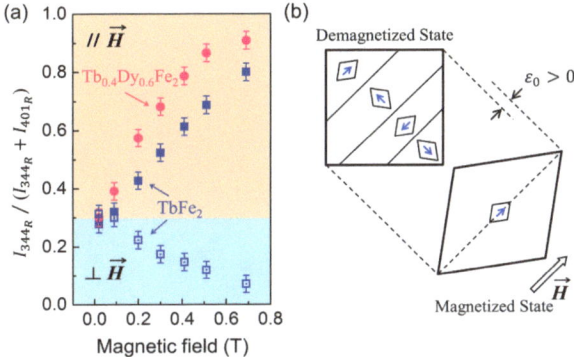

Figure 19. (a) Integrated intensity fraction of $(344)_R$ diffraction peak and (b) schematic of the microstructural evolutions in $TbFe_2$ and $Tb_{0.4}Dy_{0.6}Fe_2$ compounds under an applied magnetic field [75]. Reprinted with permission from ref. [75]. Copyright 2016 Elsevier.

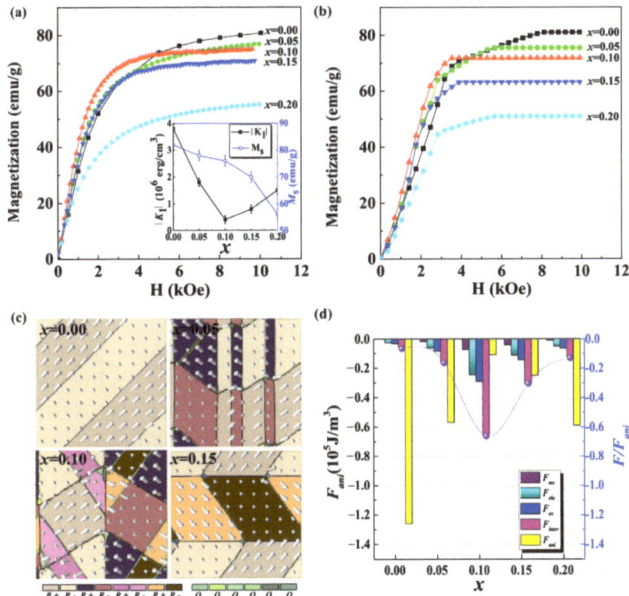

Figure 20. (a) Magnetization curves for $(Tb_{0.27}Dy_{0.73})_{1-x}(Tb_{0.1}Pr_{0.9})_x(Fe_{0.9}Co_{0.1})_2$ compounds ($0 \leq x \leq 0.2$). The inset shows the absolute values of the first-order anisotropy constant $|K_1|$ and saturation magnetization M_S at room temperature. (b) The calculated average magnetization M_{xy} from phase-field simulation for the corresponding experimental compositions. (c) Snapshots of domain structures for the samples with $0.00 \leq x \leq 0.15$. (d) Energy analysis for the samples with $0.00 \leq x \leq 0.20$ [42]. Reprinted with permission from ref. [42]. Copyright 2020 Springer Nature.

The optimum Dy content of x = 0.73 at room temperature was verified in $Tb_{1-x}Dy_xFe_2$ by first-principles calculations [78]. Combining first-principles calculations with the crystal-field approach, the critical Dy concentration was 0.78, and the corresponding magnetostrictive coefficient λ_{111} was 2700 ppm. The calculated spin-orientation diagram reproduced the experimental results for the [111] and [100] easy directions, as shown in Figure 21.

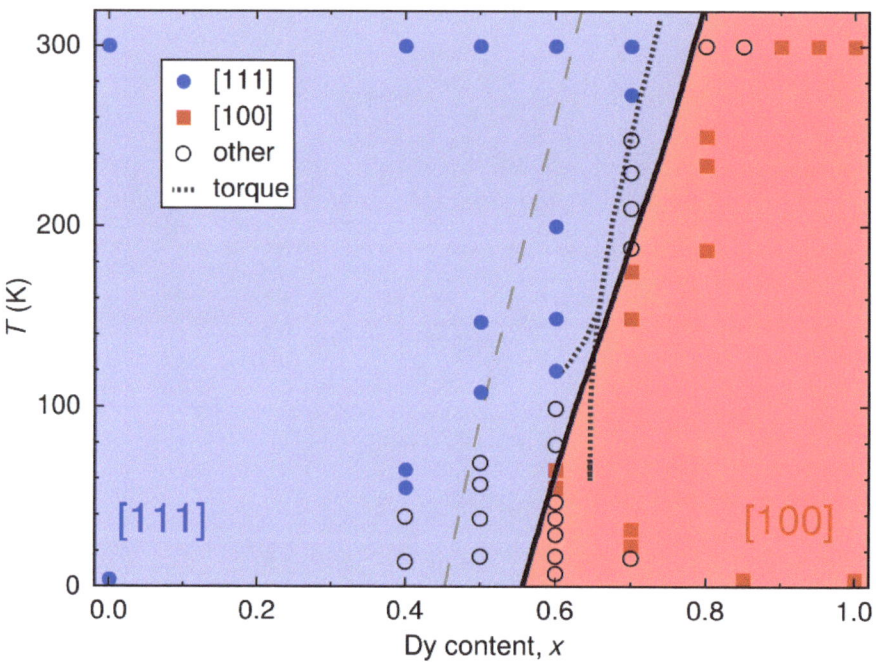

Figure 21. The easy direction of magnetization of $Tb_{1-x}Dy_xFe_2$, calculated by minimizing $E(\hat{u}, \varepsilon, x, T)$ (red- and blue-shaded regions) [78]. Reprinted with permission from ref. [78]. Copyright 2020 American Physical Society.

6. Progress on Tb-Dy-Fe Giant Magnetostrictive Composites

6.1. Polymer-Banded Tb-Dy-Fe Composite

Based on the requirements of device design, in recent years, some research has focused on giant magnetostrictive polymer composites (GMPCs) with high resistivity, a wide frequency response range and ease of formability. In a GMPC, Tb-Dy-Fe alloy particles were banded in the resin matrix in the form of particles or fibers. Tb-Dy-Fe alloy particles could be randomly dispersed (0–3 type) or arranged in a chain (1–3 type or pseudo-1–3 type) by applying different preparation methods [79]. The size, volume fraction and crystallographic orientation [80,81] of Tb-Dy-Fe alloy particles, as well as the bond strength [82] between the matrix and particles, will affect the magnetostrictive properties of the composite. After compounding, the eddy current was greatly reduced [83], which was conducive to the improvement of dynamic properties, and the mechanical properties were also greatly improved. The main characteristics of GMPCs of several Tb-Dy-Fe systems reported in the literature are summarized in Table 1.

Table 1. Properties of various GMPCs reported in the literature.

Composite	Orientation	Preparation	Magnetostrictive Particle Morphology	Magnetostrictive Particle Size	Particle Content	Magnetostriction or Comments	References
$Tb_{0.3}Dy_{0.7}Fe_{1.9}$/epoxy	<111>	8000 Oe magnetic field curing	Particles; pseudo-1–3 chain structure	>300 μm	40 vol%	1358 ppm (at 17 MPa)	[84]
$Tb_{0.3}Dy_{0.7}Fe_{1.92}$/epoxy	-	8000 Oe magnetic field curing	-	200–300 μm	40 vol%	Cut-off frequency is 6800 kHz; loss factor is only 4.3% of that for the monolithic Tb-Dy-Fe alloy (at 10 kHz and 10 mT)	[83]
$Tb_{0.4}Dy_{0.5}Nd_{0.1}(Fe_{0.8}Co_{0.2})_{1.93}$/epoxy	<111>	10 kOe magnetic field curing	Particles; pseudo-1–3 chain structure	≤150 μm	20 vol%	390 ppm (λ_a is 650 ppm at 6 kOe)	[45]
Terfenol-D/epoxy	<112>	1885 Oe magnetic field curing	Powder, particle; pseudo-1–3 chain structure	5–300 μm	70 vol%	720 ppm (at 9 MPa)	[85]
$(Tb_{0.15}Ho_{0.85}Fe_{1.9})_{0.31}$ + $(Tb_{0.3}Dy_{0.7}Fe_{1.9})_{0.69}$/epoxy	-	Pressure curing molding	Particles; pseudo-1–3 chain structure	75–180 μm	94 wt%	605 ppm	[57]
$Tb_{0.25}Dy_{0.45}Ho_{0.30}Fe_{1.9}$/epoxy	<110>	120 °C bonding molding	<110> staple fiber	0.8 mm × 0.8 mm × 12 mm	90 vol%	220 kA/M saturated magnetic field; 5 kA/M coercivity; the total loss at 20 kHz is 115 W/m³	[56]
$Tb_{0.2}Dy_{0.55}Pr_{0.25}(Fe_{0.8}Co_{0.2})_{1.93}$/epoxy	<110>	8042 Oe magnetic field curing	Particles; pseudo-1–3 chain structure	75–150 μm	30 vol%	110 ppm ($\lambda_{||}$, at 80 kA/m); 580 ppm (λ_a, at 950 kA/m)	[86]
$Tb_{0.5}Dy_{0.5}Fe_{1.95}$/epoxy	<111>	Two-step method with 10 kOe dynamic magnetic orientation	Lamellar structure	100–200 μm	57 vol%	1500 ppm	[87]
$Tb_xDy_{0.7-x}Pr_{0.3}(Fe_{0.9}B_{0.1})_{1.93}$/epoxy	<111>	8042 Oe magnetic field curing	Particles; pseudo-1–3 chain structure	60–150 μm	30 vol%	d_{33}~2.2 nm/A (Hbias~80 kA/m)	[88]

Recently, GMPCs with a layered structure were prepared by dynamic orientation in an oscillating magnetic field [89]. Jiang et al. [87] further obtained a Tb-Dy-Fe/epoxy particle composite with a high alloy particle volume fraction (57%) and high saturation magnetostriction (1500 ppm) by using a two-step dynamic orientation method, as shown in Figure 22b, and the energy density shown in Figure 22d was markedly improved. From Figure 23, it can be seen that the Tb-Dy-Fe alloy particles were first dynamically magnetically oriented in the liquid epoxy resin and then molded and concentrated in the horizontal magnetic field to remove the excess resin before curing. In addition, the defect-free matrix and anisotropic layered structure prepared by this method can effectively transfer the strain.

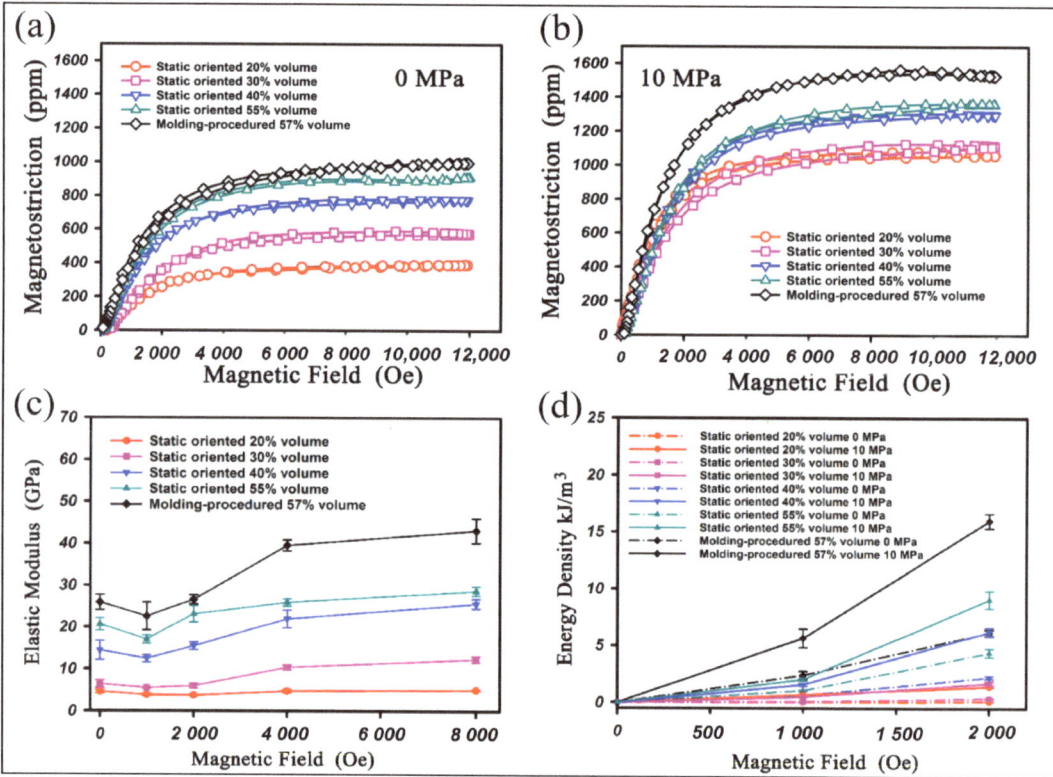

Figure 22. Magnetostriction curves of static magnetically oriented GMPCs and GMPCs subjected to a molding procedure under (**a**) 0 MPa and (**b**) 10 MPa uniaxial pressure; (**c**) elastic modulus curves and (**d**) energy densities of GMPC samples [87]. Reprinted with permission from ref. [87]. Copyright 2019 Elsevier.

6.2. Sintered Tb-Dy-Fe Material Composited with Dy-Cu Alloys

For Tb-Dy-Fe composites with polymers, a high orientation degree and a higher content of particles in the composites cannot be achieved simultaneously, which limits the further improvement of energy density.

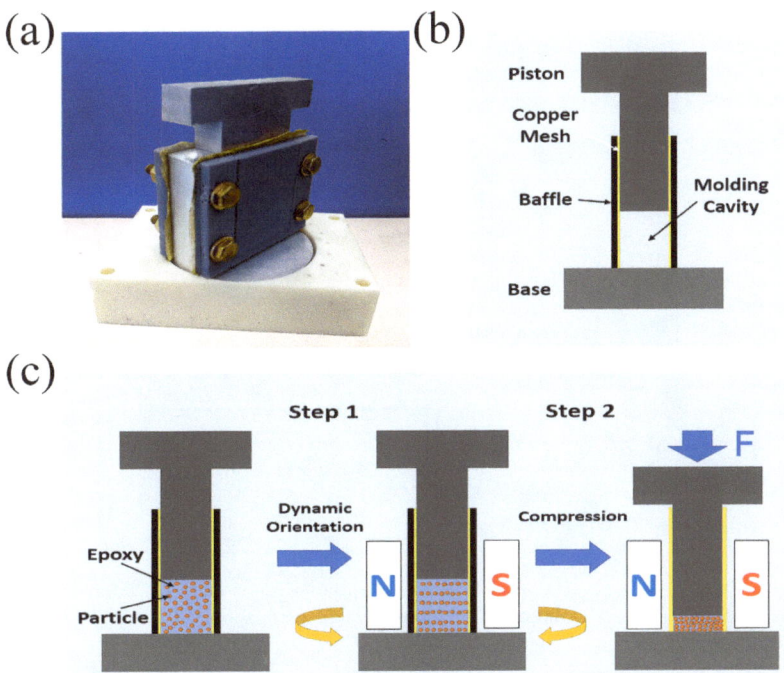

Figure 23. (**a**) The forming device; (**b**) schematic diagram of the forming device; (**c**) two-step molding procedure diagram [87]. Reprinted with permission from ref. [87]. Copyright 2019 Elsevier.

Recently, a new approach was proposed by Zhou et al. [90], which combined powder metallurgy with the magnetic field orientation. As shown in Figure 24, the powder slurry prepared with $Tb_{0.33}Dy_{0.67}Fe_{1.95}$ single-crystal particles and low-melting-point $DyCu_{1.6}$ alloy powders was wet pressed and oriented by a magnetic field. Subsequently, the compacts covered by $Dy_{1.2}Cu$ alloy thin ribbons were sintered at 1000 °C for 2 h. In this case, the $DyCu_{1.6}$ alloy powders acted as a "binder" to wet $Tb_{0.33}Dy_{0.67}Fe_{1.95}$ particles and provided a liquid channel for subsequent diffusion of $Dy_{1.2}Cu$ alloy, leading to an increase in the relative sintering density. Furthermore, a higher Tb-Dy-Fe particle content of above 90% was obtained in the sintered composites. Consequently, the <111> orientation and high sintering density effectively enhanced magnetostriction, as shown in Figure 25. More importantly, a major improvement of mechanical properties, with 176 MPa in bending strength and 71.3 MPa in tensile strength, was realized. This was attributed to the ductile Dy-Cu intergranular phase distributed along grain boundaries and the semicoherent interface between the Dy-Cu grain boundary phase and the brittle $(Tb,Dy)Fe_2$ matrix phase. The low-melting-point Dy-Cu phase was introduced as the new grain boundary phase to the Tb-Dy-Fe alloys using the sintering method, and the <111> orientation degree was improved by magnetic field orientation combined with the adjustment of the Tb/Dy ratio and particle morphology. As a result, high mechanical properties and high magnetostrictive properties were obtained in the sintered Tb-Dy-Fe/Dy-Cu composites. Mechanically, the bending strength, fracture toughness and tensile strength were respectively 3.67, 2.41 and 2.55 times those of the directionally solidified polycrystalline Tb-Dy-Fe alloy [90].

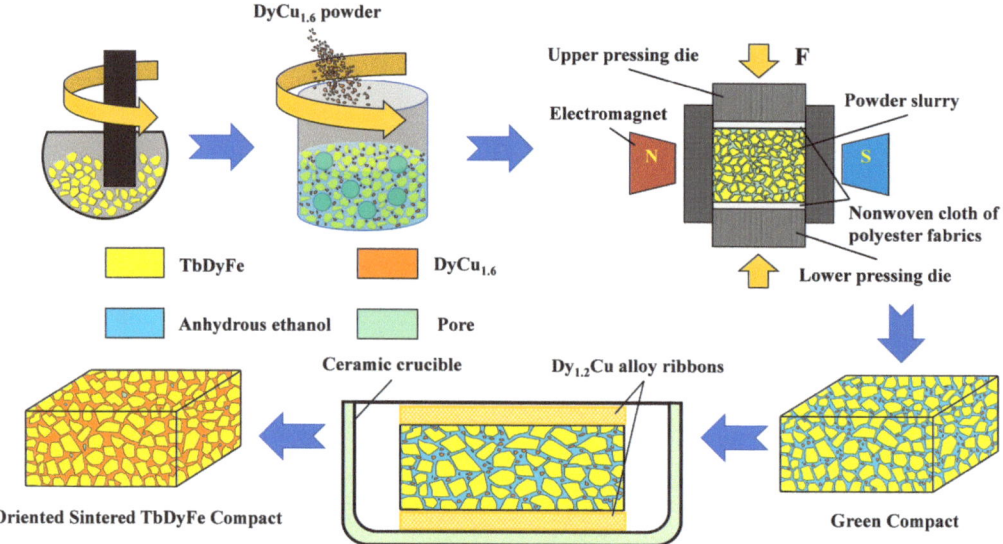

Figure 24. Schematic diagram of preparation procedure of the oriented sintered $Tb_{0.33}Dy_{0.67}Fe_{1.95}$ compacts [90]. Reprinted with permission from ref. [90]. Copyright 2022 Elsevier.

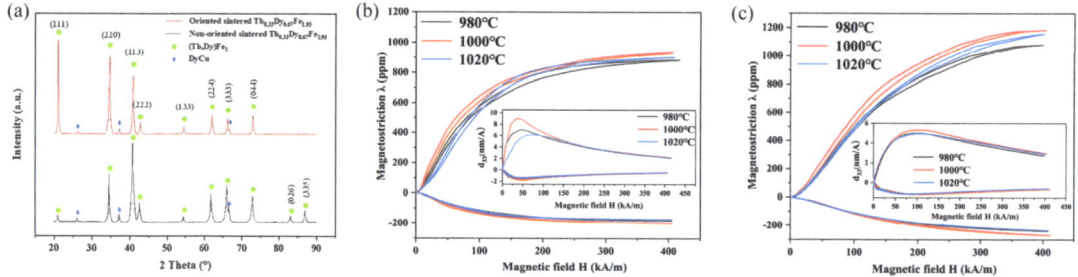

Figure 25. (**a**) XRD of the diffraction patterns of oriented and nonoriented sintered $Tb_{0.33}Dy_{0.67}Fe_{1.95}$ compacts at 1000 °C and $\lambda_{//}$, $\lambda_{\perp}$ magnetostriction curves, dependence of strain coefficient (d_{33}) of the oriented sintered $Tb_{0.33}Dy_{0.67}Fe_{1.95}$ compacts on magnetic field (H) under uniaxial pressure (**b**) 0 MPa, and (**c**) 10 MPa [90]. Reprinted with permission from ref. [90]. Copyright 2022 Elsevier.

7. Progress in Application of Tb-Dy-Fe Alloys

7.1. Tb-Dy-Fe Giant Magnetostrictive Thin Film

The research of small-scale magnetic structures is closely related to the design of microdevices and has attracted extensive attention. Tb-Dy-Fe thin film has certain applications in microelectromechanical systems (MEMS), such as microactuators and force sensors, because of its high sensitivity and large strain.

It was reported that films grown at higher substrate temperature have the combination of out-of-plane magnetic anisotropy and in-plane magnetic anisotropy [91]. Panduranga et al. obtained high-quality magnetoelastic film using sputtered Terfenol-D film after substrate heating and annealing crystallization at 450 °C [92]. In addition, another key problem of small-scale Tb-Dy-Fe materials is oxidation. It was found that the composition ratio of rare-earth oxides can be determined by the anomalous X-ray scattering method [93].

The magnetic properties of the films obtained by magnetron sputtering were similar to those of bulk single-crystal materials [92]. The magnetic properties of the micropatterned films obtained by photolithography and argon etching had little change relative to the continuous films, which was considered to be related to the oxidative passivation of the sidewall, as well as the strong dipole pair ratio of 3 μm MFM and the strong blue color in the 20 μm PEEM image in Figure 26. The results showed that it has a pseudo-single domain structure [94].

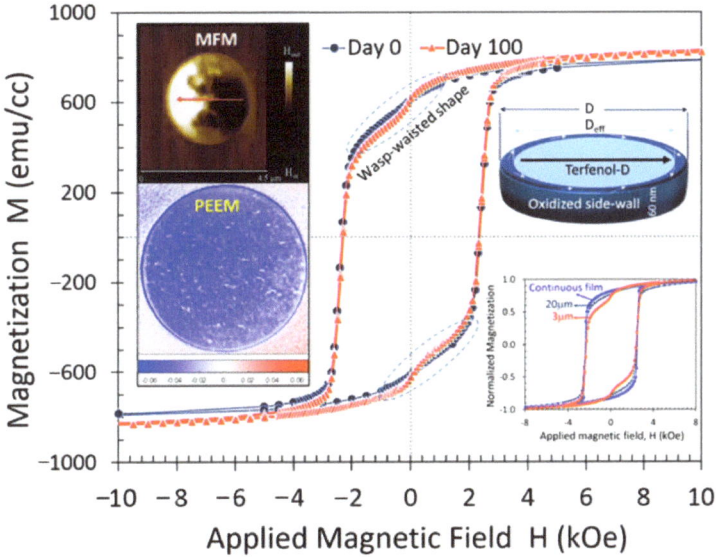

Figure 26. M vs. H plots of patterned Terfenol-D microdisks measured over the course of 100 days. Bottom-right inset shows the normalized M vs. H plots of continuous, 20 μm diameter disks and 3 μm diameter disks. Top-right inset shows the schematic of Terfenol-D disk with sidewall oxidation. Top-left inset shows magnetic force microscope image of an individual 3 μm disk upon initializing with an in-plane saturation magnetic field of 0.5 T and PEEM image of 20 μm disk indicating the single domain state when measured with Dy-M_5 edge [94]. Reprinted with permission from ref. [94]. Copyright 2021 Elsevier.

It was reported that the bimaterial cantilever structure of the magnetostrictive film prepared by electrodeposition had a large magnetostrictive coefficient of about 1250 ppm under an 11 kOe magnetic field (Figure 27), and the energy density was able to reach 100–165 kJ/m^3 [95]. In microdevices with a cantilever structure, a multilayer structure was generally used for better device performance. For instance, Tb-Dy-Fe/graphene/Tb-Dy-Fe multilayer film was used to replace the traditional three-layer Tb-Dy-Fe film in recent research, which resulted in a reduction in the dynamic response delay time.

7.2. Application in Microsensors and Other Devices

Magnetostrictive materials are usually used in actuators and sonar transducers. Recently, Tb-Dy-Fe materials were used in various high-sensitivity magnetostrictive sensors, such as current sensors [96,97], magnetic sensors [98] and torque sensors [99]. To explore its application, some researchers sputtered Tb-Dy-Fe film on a Fe-Co substrate, and the composite film was expected to be used in high-precision nondestructive testing [100]. In recent years, an application in wireless temperature measurement was developed, in which Terfenol-D was used to increase the temperature coefficient of resonant frequency [101].

A Tb-Dy-Fe magnetostrictive transducer combined with an optical fiber sensor is able to improve the accuracy of magnetic field detection. A series of high-precision magnetic field sensors were prepared by combining them with a fiber Bragg grating (FBG) sensor [102–104] and phase-shifted fiber Bragg grating (PS-FBG) [105]. The high-sensitivity response of the device was improved by the design of different systems. Feng et al. prepared a high-finesse fiber-optic extrinsic Fabry–Perot interferometer (EFPI)-based sensor, and its sensitivity was significantly improved based on the design of a mechanical amplification structure [106].

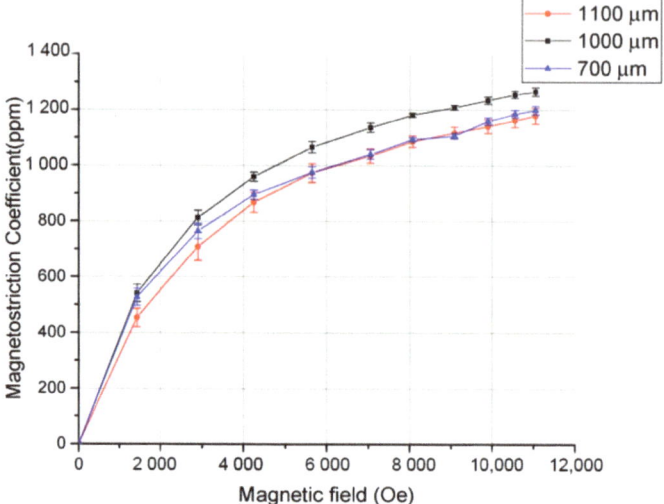

Figure 27. Magnetostriction coefficients of the $Tb_{0.36}Dy_{0.64}Fe_{1.9}$ film obtained from three cantilevers with lengths 700, 1000 and 1100 μm [95]. Reprinted from ref. [94].

8. Summary and Prospects

Improvements in the directional solidification preparation process, such as directional solidification with a strong magnetic field, are effective ways to obtain better orientation for directionally solidified Tb-Dy-Fe alloys. Tb-Dy-Fe composites prepared by the resin bonding method or low-melting-point alloy sintering method are more based on the improvement of high-frequency application properties and mechanical properties. At present, through the improvement of composite methods such as dynamic magnetic field orientation, the volume fraction and properties of Tb-Dy-Fe alloy particles have been greatly improved. In order to achieve better practical value, we need to constantly explore Tb-Dy-Fe series alloys with large magnetostriction, high mechanical properties, low loss and low cost. It is worth mentioning that the Dy-Cu phase was introduced as a new grain boundary phase to Tb-Dy-Fe alloys by the sintering method, and the <111> orientation was improved by the magnetic field orientation combined with the adjustment of the Tb/Dy ratio and particle morphology. High mechanical and magnetostrictive properties were achieved in sintered Tb-Dy-Fe/Dy-Cu composites. Meanwhile, sintered composite materials based on the reconstruction of the grain boundary phase also provide new ideas for the development of Tb-Dy-Fe materials with excellent comprehensive properties, including high magnetostriction, high mechanical properties, high corrosion resistance and high resistivity.

The studies of phase structure and ferromagnetic MPB, combined with domain structure, phase-field simulation and first-principles calculations, are very helpful to deeply understand the magnetostriction origin of Tb-Dy-Fe alloys. At the same time, such studies also facilitate an understanding of the influence of element substitution or developing a new alloy system. However, the mechanism and theoretical system still need more in-depth research to provide further support.

The research progress on the application of Tb-Dy-Fe alloys is extensive, especially in microdevices and various high-sensitivity sensors based on Tb-Dy-Fe films. However, bulk Tb-Dy-Fe materials could play a crucial role in a broader field on the basis of the effective improvement of comprehensive properties.

Author Contributions: Conceptualization, Z.Y. and J.L.; methodology, Z.Y. and J.L.; validation, Z.Y. and J.L.; formal analysis, Z.Y. and J.L.; investigation, Z.Y. and J.L.; resources, Z.Y. and J.L.; data curation, Z.Y. and J.L.; writing—original draft preparation, Z.Y. and J.L.; writing—review and editing, Z.Y., J.L., Z.Z. and J.G.; visualization, Z.Y. and J.L.; supervision, J.L. and X.G.; project administration, X.G.; funding acquisition, X.B. and X.G. All authors have read and agreed to the published version of the manuscript.

Funding: This research was funded by the National Key R & D Program of China, grant number 2021YFB3501403; the State Key Laboratory for Advanced Metals and Materials, grant numbers 2017Z-11 and 2018Z-07; and Fundamental Research Funds for the Central Universities, grant numbers FRF-GF-17-B2, FRF-GF-19-028B and FRF-GF-20-23B.

Institutional Review Board Statement: Not applicable.

Informed Consent Statement: Not applicable.

Data Availability Statement: Data available in a publicly accessible repository.

Acknowledgments: This work is supported by the National Key R & D Program of China (grant no. 2021YFB3501403), the State Key Laboratory for Advanced Metals and Materials (2017Z-11, 2018Z-07), and the Fundamental Research Funds for the Central Universities (FRF-GF-17-B2, FRF-GF-19-028B, and FRF-GF-20-23B).

Conflicts of Interest: The authors declare no conflict of interest.

References

1. Clark, A.E. Magnetic and Magnetoelastic Properties of Highly Magnetostrictive Rare Earth-Iron Laves Phase Compounds. *AIP Conf. Proc.* **1974**, *18*, 1015–1029. [CrossRef]
2. Clark, A.; Crowder, D. High temperature magnetostriction of $TbFe_2$ and $Tb_{0.27}Dy_{0.73}Fe_2$. *IEEE Trans. Magn.* **1985**, *21*, 1945–1947. [CrossRef]
3. Zhou, S.Z.; Gao, X.X. *Magnetostrictive Materials*; Metallurgical Industry Press: Beijing, China, 2017.
4. Zhou, S.Z.; Zhao, Q.; Zhang, M.C.; Gao, X.X.; Wang, Z.C.; Shi, Z.H. Giant magnetostrictive materials of Tb-Dy-Fe alloy with [110] axial alignment. *Prog. Nat. Sci.* **1998**, *6*, 83–86.
5. Ma, T.; Jiang, C.; Xiao, F.; Xu, H. Magnetostriction of $Tb_{0.36}Dy_{0.64}(Fe_{1-x}Co_x)_2$ (x = 0-0.20) <112>-oriented crystals. *J. Alloys Compd.* **2006**, *414*, 276–281. [CrossRef]
6. Wu, W.; Zhang, M.C.; Gao, X.X.; Zhou, S.Z. Effect of two-steps heat treatment on the mechanical properties and magnetostriction of <110> oriented TbDyFe giant magnetostrictive material. *J. Alloys Compd.* **2006**, *416*, 256–260. [CrossRef]
7. Ren, W.J.; Zhang, Z.D. Progress in bulk $MgCu_2$-type rare-earth iron magnetostrictive compounds. *Chin. Phys. B* **2013**, *22*, 077507. [CrossRef]
8. Liu, J.H.; Zhang, T.L.; Wang, J.M.; Jiang, C.B. Giant Magnetostrictive Materials and Their Applications. *Mater. China* **2012**, *31*, 1–12.
9. Pan, Z.B.; Liu, J.J.; Liu, X.Y.; Wang, J.; Du, J.; Si, P.Z. Structural, magnetic and magnetostrictive properties of Laves-phase compounds $Tb_xHo_{0.9-x}Nd_{0.1}Fe_{1.93}$ ($0 \leq x \leq 0.40$). *Mater. Chem. Phys.* **2014**, *148*, 82–86. [CrossRef]
10. Wun-Fogle, M.; Restorff, J.B.; Clark, A.E. Hysteresis and magnetostriction of $Tb_xDy_yHo_{1-x-y}Fe_{1.95}$ [112] dendritic rods. *J. Appl. Phys.* **1999**, *85*, 6253–6255. [CrossRef]
11. Guo, Z.J.; Busbridge, S.C.; Wang, B.W.; Zhang, Z.D.; Zhao, X.G. Structure and magnetic and magnetostrictive properties of $(Tb_{0.7}Dy_{0.3})_{0.7}Pr_{0.3}(Fe_{1-x}Co_x)_{1.85}$ ($0 \leq x \leq 0.6$). *IEEE Trans. Magn.* **2001**, *37*, 3025–3027. [CrossRef]
12. Xu, L.H.; Jiang, C.B.; Xu, H.B. Magnetostriction and electrical resistivity of Si doped $Tb_{0.3}Dy_{0.7}Fe_{1.95}$ oriented crystals. *Appl. Phys. Lett.* **2006**, *89*, 192507. [CrossRef]
13. Clark, A.E. Chapter 7 Magnetostrictive Rare Earth-Fe_2 Compounds. In *Handbook of Ferromagnetic Materials*; Elsevier: Amsterdam, The Netherlands, 1980; Volume 1, pp. 531–589.
14. Yang, S.; Bao, H.X.; Zhou, C.; Wang, Y.; Ren, X.B.; Song, X.P.; Matsushita, Y.; Katsuya, Y.; Tanaka, M.; Kobayashi, K. ChemInform Abstract: Structural Changes Concurrent with Ferromagnetic Transition. *Chin. Phys. B* **2013**, *22*, 046101. [CrossRef]
15. Wei, S.R.; Song, X.P.; Yang, S.; Deng, J.K.; Wang, Y. Monte Carlo Simulation on the Magnetization Rotation near Magnetic Morphotropic Phase Boundary; SPIE: Shenzhen, China, 2012; Volume 8409. [CrossRef]
16. Yang, S.; Bao, H.; Chao, Z.; Yu, W.; Gao, J. Large Magnetostriction from Morphotropic Phase Boundary in Ferromagnets. *Phys. Rev. Lett.* **2010**, *104*, 197201. [CrossRef]
17. Fohtung, E. Magnetostriction Fundamentals. In *Encyclopedia of Smart Materials*; Rensselaer Polytechnic Institute: Troy, NY, USA, 2022; pp. 130–133. [CrossRef]
18. Chang, T.; Zhou, C.; Chang, K.; Wang, B.; Shi, Q.; Chen, K.; Chen, Y.-S.; Ren, Y.; Yang, S. Local structure study on magnetostrictive material $Tb_{1-x}Dy_xFe_2$. *J. Appl. Phys.* **2020**, *127*, 235102. [CrossRef]

19. Kang, D.Z.; Liu, J.H.; Jiang, C.B.; Xu, H.B. Control of solid-liquid interface morphology and radial composition distribution: TbDyFe single crystal growth. *J. Alloys Compd.* **2015**, *621*, 331–338. [CrossRef]
20. Palit, M.; Banumathy, S.; Singh, A.K.; Pandian, S.; Chattopadhyay, K. Crystallography of solid-liquid interface and evolution of texture during directional solidification of $Tb_{0.3}Dy_{0.7}Fe_{1.95}$ alloy. *Intermetallics* **2011**, *19*, 357–368. [CrossRef]
21. Palit, M.; Banumathy, S.; Singh, A.K.; Pandian, S.; Chattopadhyay, K. Orientation Selection and Microstructural Evolution in Directionally Solidified $Tb_{0.3}Dy_{0.7}Fe_{1.95}$. *Metall. Mater. Trans. A* **2016**, *47*, 1729–1739. [CrossRef]
22. Kang, D.Z.; Liu, J.H.; Jiang, C.B.; Xu, H.B. Correlation between Growth Twinning and Crystalline Reorientation of Faceted Growth Materials during Directional Solidification. *Cryst. Growth Des.* **2015**, *15*, 3092–3095. [CrossRef]
23. Jiang, C.B.; Zhou, S.S.; Zhang, M.C.; Run, W. The preferred orientation, microstructure and magnetostriction in directionally solidified TbDyFe alloys. *Acta Metall. Sin.* **1998**, *34*, 164–170.
24. Kang, D.Z.; Zhang, T.L.; Jiang, C.B.; Xu, H.B. Preferred orientation transition mechanism of faceted-growth materials with FCC structure: Competitive advantage depends on 3D microstructure morphologies. *J. Alloys Compd.* **2018**, *741*, 14–20. [CrossRef]
25. Sun, Z.H.I.; Guo, M.; Vleugels, J.; Van der Biest, O.; Blanpain, B. Strong static magnetic field processing of metallic materials: A review. *Curr. Opin. Solid State Mater. Sci.* **2012**, *16*, 254–267. [CrossRef]
26. Wang, J.; Fautrelle, Y.; Ren, Z.M.; Li, X.; Nguyen-Thi, H.; Mangelinck-Noel, N.; Salloum, A.; Zhong, Y.B.; Kaldre, I.; Bojarevics, A. Thermoelectric magnetic force acting on the solid during directional solidification under a static magnetic field. *Appl. Phys. Lett.* **2012**, *101*, 1331–1333.
27. Gao, P.F.; Liu, T.; Dong, M.; Yuan, Y.; Wang, Q. Magnetic domain structure, crystal orientation, and magnetostriction of $Tb_{0.27}Dy_{0.73}Fe_{1.95}$ solidified in various high magnetic fields. *J. Magn. Magn. Mater.* **2016**, *401*, 755–759. [CrossRef]
28. Gao, P.F.; Liu, T.; Chai, S.W.; Dong, M.; Wang, Q. Influence of magnetic flux density and cooling rate on orientation behavior of $Tb_{0.27}Dy_{0.73}Fe_{1.95}$ alloy during solidification process. *Acta Phys. Sin. Chin. Ed.* **2016**, *65*, 335–342. [CrossRef]
29. Dong, S.L.; Liu, T.; Dong, M.; Guo, X.Y.; Yuan, S.; Wang, Q. Enhanced magnetostriction of Tb–Dy–Fe via simultaneous <111>-crystallographic orientation and -morphological alignment induced by directional solidification in high magnetic fields. *Appl. Phys. Lett.* **2020**, *116*, 053903. [CrossRef]
30. Wang, Q.; Liu, T.; Wang, K.; Gao, P.F.; Liu, Y.; He, J.C. Progress on High Magnetic Field-Controlled Transport Phenomena and Their Effects on Solidification Microstructure. *ISIJ Int.* **2014**, *54*, 516–525. [CrossRef]
31. Dong, M.; Liu, T.; Guo, X.Y.; Liu, Y.R.; Dong, S.L.; Wang, Q. Enhancement of mechanical properties of $Tb_{0.27}Dy_{0.73}Fe_{1.95}$ alloy by directional solidification in high magnetic field. *Mat. Sci. Eng. A Struct.* **2020**, *785*, 139377. [CrossRef]
32. Gao, P.F.; Liu, T.; Dong, M.; Yuan, Y.; Wang, K.; Wang, Q. Magnetostrictive gradient in $Tb_{0.27}Dy_{0.73}Fe_{1.95}$ induced by high magnetic field gradient applied during solidification. *Funct. Mater. Lett.* **2016**, *09*, 1650003. [CrossRef]
33. Gao, P.F. Evolution of Microstructure and Magnetostrictive Performance of $Tb_{0.27}Dy_{0.73}Fe_{1.95}$ Alloy Solidified in High Magnetic Field Gradients. Ph.D. Dissertation, Northeastern University, Boston, MA, USA, 2015.
34. Liu, T.; Gao, P.F.; Dong, M.; Xiao, Y.B.; Wang, Q. Effect of cooling rate on magnetostriction gradients of $Tb_{0.27}Dy_{0.73}Fe_{1.95}$ alloys solidified in high magnetic field gradients. *AIP Adv.* **2016**, *6*, 056216. [CrossRef]
35. Ren, W.J.; Liu, J.J.; Li, D.; Liu, W.; Zhao, X.G.; Zhang, Z.D. Direct experimental evidence for anisotropy compensation between Dy^{3+} and Pr^{3+} ions. *Appl. Phys. Lett.* **2006**, *89*, 122506. [CrossRef]
36. Shi, Y.G.; Chen, Z.Y.; Wang, L.; Hu, C.C.; Pan, Q.; Shi, D.N. Synthesis and magnetostrictive properties of $Pr_{1-x}Dy_x(Fe_{0.8}Co_{0.2})_{1.93}$ cubic Laves compounds. *AIP Adv.* **2016**, *6*, 056207. [CrossRef]
37. Zhang, G.B.; Zheng, W.G.; Cui, Y.; Shi, Y.G.; Shi, D.N. Structural, Magnetic, and Magentostrictive Properties of $Dy_{1-x}(Tb_{0.2}Pr_{0.8})_xFe_{1.93}$ ($0 \leq x \leq 0.5$) Compounds. *J. Supercond. Novel Magn.* **2017**, *31*, 2217–2220. [CrossRef]
38. Shi, Y.G.; Tang, S.L.; Wang, R.L.; Su, H.L.; Han, Z.D.; Lv, L.Y.; Du, Y.W. High-pressure synthesis of giant magnetostrictive $Pr_xTb_{1-x}Fe_{1.9}$ alloys. *Appl. Phys. Lett.* **2006**, *89*, 202503. [CrossRef]
39. Shi, Y.G.; Tang, S.L.; Zhai, L.; Huang, H.B.; Wang, R.L.; Yu, J.Y.; Du, Y.W. Composition anisotropy compensation and magnetostriction in $Pr(Fe_{1-x}Co_x)_{1.9}$ ($0 \leq x \leq 0.5$) cubic Laves alloys. *Appl. Phys. Lett.* **2008**, *92*, 212507. [CrossRef]
40. Shi, Y.G.; Tang, S.L.; Lv, L.Y.; Fan, J.Y. Magnetic and magnetostrictive properties in high-pressure synthesized $Dy_{1-x}Pr_xFe_{1.9}$ ($0 \leq x \leq 1$) cubic Laves alloys. *J. Alloys Compd.* **2010**, *506*, 533–536. [CrossRef]
41. Zhang, G.B.; Liu, Y.D.; Kan, C.X.; Shi, Y.G.; Shi, D.N. Structure and magnetostriction in $(Tb_{0.2}Pr_{0.8})_xDy_{1-x}Fe_{1.93}$ Laves compounds synthesised by high-pressure annealing. *Mater. Res. Bull.* **2019**, *112*, 174–177. [CrossRef]
42. Hu, C.C.; Zhang, Z.; Cheng, X.X.; Huang, H.B.; Shi, Y.G.; Chen, L.Q. Ultrasensitive magnetostrictive responses at the pre-transitional rhombohedral side of ferromagnetic morphotropic phase boundary. *J. Mater. Sci.* **2020**, *56*, 1713–1729. [CrossRef]
43. Hu, C.C.; Shi, Y.G.; Shi, D.N.; Zhou, X.G.; Fan, J.Y.; Lv, L.Y.; Tang, S.L. Optimization on magnetic transitions and magnetostriction in $Tb_xDy_yNd_z(Fe_{0.9}Co_{0.1})_{1.93}$ compounds. *J. Appl. Phys.* **2013**, *114*, 143906. [CrossRef]
44. Liu, J.J.; Pan, Z.B.; Liu, X.Y.; Zhang, Z.R.; Song, X.H.; Ren, W.J. Large magnetostriction and direct experimental evidence for anisotropy compensation in $Tb_{0.4-x}Nd_xDy_{0.6}(Fe_{0.8}Co_{0.2})_{1.93}$ Laves compounds. *Mater. Lett.* **2014**, *137*, 274–276. [CrossRef]
45. Yin, H.Y.; Liu, J.J.; Pan, Z.B.; Liu, X.Y.; Liu, X.C.; Liu, L.D.; Du, J.; Si, P.Z. Magnetostriction of $Tb_xDy_{0.9-x}Nd_{0.1}(Fe_{0.8}Co_{0.2})_{1.93}$ compounds and their composites ($0.20 \leq x \leq 0.60$). *J. Alloys Compd.* **2014**, *582*, 583–587. [CrossRef]
46. Shi, Y.G.; Tang, S.L.; Huang, Y.J.; Lv, L.Y.; Du, Y.W. Anisotropy compensation and magnetostriction in $Tb_xNd_{1-x}Fe_{1.9}$ cubic Laves alloys. *Appl. Phys. Lett.* **2007**, *90*, 142515. [CrossRef]

47. Jammalamadaka, S.N.; Markandeyulu, G.; Balasubramaniam, K. Magnetostriction and anisotropy compensation in Tb$_x$Dy$_{0.9-x}$Nd$_{0.1}$Fe$_{1.93}$ (0.2 ≤ x ≤ 0.4). *Appl. Phys. Lett.* **2010**, *97*, 242502. [CrossRef]
48. Pan, Z.B.; Liu, J.J.; Liu, X.Y.; Li, X.; Song, X.H.; Zhang, Z.R.; Ren, W.J. Structural, magnetic and magnetoelastic properties of Laves-phase Tb$_{0.3}$Dy$_{0.6}$Nd$_{0.1}$(Fe$_{1-x}$Co$_x$)$_{1.93}$ compounds (0 ≤ x ≤ 0.40). *Intermetallics* **2015**, *64*, 1–5. [CrossRef]
49. Chen, Z.Y.; Shi, Y.G.; Wang, L.; Pan, Q.; Li, H.F.; Shi, D.N. Structure and magnetic properties of melt-spun Tb$_{0.2}$Nd$_{0.8}$(Fe$_{0.8}$Co$_{0.2}$)$_{1.9}$ compound. *J. Alloys Compd.* **2016**, *656*, 259–262. [CrossRef]
50. Wang, L. Magnetostrictive Properties in (Pr,Nd)Modulated Laves Compounds. Master Dissertation, Nanjing University of Aeronautics and Astronautics, Nanjing, China, 2018.
51. Shi, Y.G.; Wang, L.; Zheng, W.G.; Chen, Z.Y.; Shi, D.N. Effects of Co substitution for Fe on the structural and magnetostrictive properties of melt-spun Tb$_{0.2}$Nd$_{0.8}$Fe$_{1.9}$ ribbons. *J. Magn. Magn. Mater.* **2017**, *433*, 116–119. [CrossRef]
52. Murtaza, A.; Li, Y.B.; Mi, J.W.; Zuo, W.L.; Ghani, A.; Dai, Z.Y.; Yao, K.K.; Hao, C.X.; Yaseen, M.; Saeed, A. Spin configuration, magnetic and magnetostrictive properties of Tb$_{0.27}$Dy$_{0.73-x}$Nd$_x$Fe$_2$ compounds. *Mater. Chem. Phys.* **2020**, *249*, 122951. [CrossRef]
53. Wang, B.; Lv, Y.; Li, G.; Huang, W.M.; Sun, Y.; Cui, B.Z. The magnetostriction and its ratio to hysteresis for Tb-Dy-Ho-Fe alloys. *J. Appl. Phys.* **2014**, *115*, 7282. [CrossRef]
54. Wang, B.W.; Cao, S.Y.; Huang, W.M.; Sun, Y.; Weng, L.; Zhao, Z.Z. Phase Relationship and Magnetostriction of Tb-Dy-Ho-Fe Alloys. *IEEE Trans. Appl. Supercond.* **2016**, *26*, 1–4. [CrossRef]
55. Pan, Z.B.; Liu, J.J.; Si, P.Z.; Ren, W.J. Magnetostriction of Laves Tb$_{0.1}$Ho$_{0.9-x}$Pr$_x$ (Fe$_{0.8}$Co$_{0.2}$)$_{1.93}$ alloys. *Mater. Res. Bull.* **2016**, *77*, 122–125. [CrossRef]
56. Zhao, R.; Wang, B.M.; Huang, W.M.; Yan, J.W. High frequency magnetic properties of polymer-bonded Tb-Dy-Ho-Fe fiber composites. *Ferroelectrics* **2018**, *530*, 51–59. [CrossRef]
57. Zhao, R.; Wang, B.W.; Cao, S.Y.; Xiao, J. Effect of Ho Doping and Annealing on Magnetostrictive Properties of Tb-Dy-Ho-Fe/Epoxy Composites. *Chem. Eng. Trans.* **2016**, *55*, 301–306. [CrossRef]
58. Wang, N.J.; Liu, Y.; Zhang, H.W.; Chen, X.; Li, Y.X. Fabrication, magnetostriction properties and applications of Tb-Dy-Fe alloys: A review. *China Foundry* **2016**, *13*, 75–84. [CrossRef]
59. Wang, N.J.; Liu, Y.; Zhang, H.W.; Chen, X.; Li, Y.X. Effect of Co, Cu, Nb, Ti, V on magnetostriction and mechanical properties of TbDyFe alloys. *Intermetallics* **2018**, *100*, 188–192. [CrossRef]
60. Jiang, C.B.; Ma, T.Y.; Xu, H.B. A kind of wide operating temperature range giant magnetostrictive alloys. *J. Alloys Compd.* **2008**, *449*, 156–160. [CrossRef]
61. Ma, T.Y.; Jiang, C.B.; Xu, X.; Zhang, H.; Xu, H.B. The Co-doped Tb$_{0.36}$Dy$_{0.64}$Fe$_2$ magnetostrictive alloys with a wide operating temperature range. *J. Magn. Magn. Mater.* **2005**, *292*, 317–324. [CrossRef]
62. Zhou, C.; Zeng, Y.Y.; Chang, T.Y.; Wang, Y.; Zhang, A.Z.; Liu, S.Y.; Tian, F.H.; Zhang, Y.; Song, X.P.; Yang, S. Ferromagnetic and magnetostrictive properties of Tb$_{0.3}$Dy$_{0.7}$(Co$_{1-x}$Fe$_x$)$_2$ alloys. *Jpn. J. Appl. Phys.* **2019**, *58*, 050921. [CrossRef]
63. Westwood, P.; Abell, J.S.; Pitman, K.C. Phase relationships in the Tb-Dy-Fe ternary system. *J. Appl. Phys.* **1990**, *67*, 4998–5000. [CrossRef]
64. Bi, Y.J.; Abell, J.S.; Hwang, A.M.H. Defects in Terfenol-D crystals. *J. Magn. Magn. Mater.* **1991**, *99*, 159–166. [CrossRef]
65. Peterson, D.T.; Verhoeven, J.D.; McMasters, O.D.; Spitzig, W.A. Strength of Terfenol-D. *J. Appl. Phys.* **1989**, *65*, 3712–3713. [CrossRef]
66. Wu, W.; Tang, H.J.; Zhang, M.C.; Gao, X.X.; He, J.P.; Zhou, S.Z. Effect of heat treatment on the mechanical properties of <110> oriented TbDyFe giant magnetostrictive material. *J. Alloys Compd.* **2006**, *413*, 96–100. [CrossRef]
67. Wang, N.J.; Liu, Y.; Zhang, H.W.; Chen, X.; Li, Y.X. Effect of Nb on magnetic and mechanical properties of TbDyFe alloys. *J. Magn. Magn. Mater.* **2018**, *449*, 223–227. [CrossRef]
68. Wang, N.J.; Liu, Y.; Zhang, H.W.; Chen, X.; Li, Y.X. Effect of copper on magnetostriction and mechanical properties of TbDyFe alloys. *J. Rare Earths* **2019**, *37*, 74–79. [CrossRef]
69. Zhou, Z.G.; Li, J.H.; Bao, X.Q.; Zhou, Y.Y.; Gao, X.X. Improvement of bending strength via introduced (Dy,Tb)Cu phase at grain boundary on giant magnetostrictive Tb-Dy-Fe alloy by diffusing Dy-Cu alloys. *J. Alloys Compd.* **2020**, *826*, 153959. [CrossRef]
70. Yang, S.; Ren, X. Noncubic crystallographic symmetry of a cubic ferromagnet: Simultaneous structural change at the ferromagnetic transition. *Phys. Rev. B* **2008**, *77*, 014407. [CrossRef]
71. Bergstrom, R., Jr.; Wuttig, M.; Cullen, J.; Zavalij, P.; Briber, R.; Dennis, C.; Garlea, V.O.; Laver, M. Morphotropic phase boundaries in ferromagnets: Tb$_{1-x}$Dy$_x$Fe$_2$ alloys. *Phys. Rev. Lett* **2013**, *111*, 017203. [CrossRef] [PubMed]
72. Wei, S.; Liao, X.; Gao, Y.; Yang, S.; Wang, D.; Song, X. Simulation study on exchange interaction and unique magnetization near ferromagnetic morphotropic phase boundary. *J. Phys. Condens. Matter* **2017**, *29*, 445802. [CrossRef]
73. Atzmony, U.; Dariel, M.P.; Dublon, G. Spin-orientation diagram of the pseudobinary Tb$_{1-x}$Dy$_x$Fe$_2$ Laves compounds. *Phys. Rev. B* **1977**, *15*, 3565–3566. [CrossRef]
74. Ma, T.Y.; Liu, X.L.; Pan, X.W.; Li, X.; Jiang, Y.Z.; Yan, M.; Li, H.Y.; Fang, M.X.; Ren, X.B. Local rhombohedral symmetry in Tb$_{0.3}$Dy$_{0.7}$Fe$_2$ near the morphotropic phase boundary. *Appl. Phys. Lett.* **2014**, *105*, 192407. [CrossRef]
75. Nie, Z.H.; Yang, S.; Wang, Y.D.; Wang, Z.L.; Liu, D.M.; Ren, Y.; Chang, T.T.; Zhang, R. In-situ studies of low-field large magnetostriction in Tb$_{1-x}$Dy$_x$Fe$_2$ compounds by synchrotron-based high-energy x-ray diffraction. *J. Alloys Compd.* **2016**, *658*, 372–376. [CrossRef]

76. Hu, C.C.; Yang, T.N.; Huang, H.B.; Hu, J.M.; Wang, J.J.; Shi, Y.G.; Shi, D.N.; Chen, L.Q. Phase-field simulation of domain structures and magnetostrictive response in $Tb_{1-x}Dy_xFe_2$ alloys near morphotropic phase boundary. *Appl. Phys. Lett.* **2016**, *108*, 141908. [CrossRef]
77. Hu, C.C.; Zhang, Z.; Yang, T.N.; Li, W.; Chen, L.Q. Phase-field simulation of magnetic microstructure and domain switching in $(Tb_{0.27}Dy_{0.73})Fe_2$ single crystal. *AIP Adv.* **2021**, *11*, 015207. [CrossRef]
78. Patrick, C.E.; Marchant, G.A.; Staunton, J.B. Spin Orientation and Magnetostriction of $Tb_{1-x}Dy_xFe_2$ from First Principles. *Phys. Rev. Appl.* **2020**, *14*, 014091. [CrossRef]
79. Elhajjar, R.; Law, C.T.; Pegoretti, A. Magnetostrictive polymer composites: Recent advances in materials, structures and properties. *Prog. Mater. Sci.* **2018**, *97*, 204–229. [CrossRef]
80. Lo, C.Y.; Or, S.W.; Chan, H.L.W. Large Magnetostriction in Epoxy-Bonded Terfenol-D Continuous-Fiber Composite with [112] Crystallographic Orientation. *IEEE Trans. Magn.* **2006**, *42*, 3111–3113. [CrossRef]
81. Ho, K.K.; Henry, C.P.; Altin, G.; Carman, G.P. Crystallographically Aligned Terfenol-D/Polymer Composites for a Hybrid Sonar Device. *Integr. Ferroelectr.* **2006**, *83*, 121–138. [CrossRef]
82. Dong, X.F.; Qi, M.; Guan, X.C.; Li, J.H.; Ou, J.P. Magnetostrictive properties of titanate coupling agent treated Terfenol-D composites. *J. Magn. Magn. Mater.* **2012**, *324*, 1205–1208. [CrossRef]
83. Meng, H.; Zhang, T.L.; Jiang, C.B. Cut-off frequency of magnetostrictive materials based on permeability spectra. *J. Magn. Magn. Mater.* **2012**, *324*, 1933–1937. [CrossRef]
84. Meng, H.; Zhang, T.L.; Jiang, C.B.; Xu, H.B. Grain-<111>-oriented anisotropy in the bonded giant magnetostrictive material. *Appl. Phys. Lett.* **2010**, *96*, 102501. [CrossRef]
85. Kaleta, J.; Lewandowski, D.; Mech, R. Magnetostriction of field-structural composite with Terfenol-D particles. *Arch. Civ. Mech. Eng.* **2015**, *15*, 897–902. [CrossRef]
86. Lin, L.L.; Liu, J.J.; Shen, W.C.; Ding, Q.L.; Wang, M.K.; Du, J.; Si, P.Z. Magnetomechanical behavior of $Tb_{0.2}Dy_{0.8-x}Pr_x(Fe_{0.8}Co_{0.2})_{1.93}$/epoxy pseudo-1-3 particulate composites. *Appl. Phys. A* **2018**, *124*, 706. [CrossRef]
87. Li, B.C.; Zhang, T.L.; Wu, Y.Y.; Jiang, C.B. High-performance magnetostrictive composites with large particles volume fraction. *J. Alloys Compd.* **2019**, *805*, 1266–1270. [CrossRef]
88. Lv, X.; Liu, J.; Ding, Q.; Wang, M.; Pan, Z. Textured Orientation and Dynamic Magnetoelastic Properties of Epoxy-Based $Tb_xDy_{0.7-x}Pr_{0.3}(Fe_{0.9}B_{0.1})_{1.93}$ Particulate Composites. *J. Supercond. Novel Magn.* **2020**, *33*, 3857–3864. [CrossRef]
89. Du, T.; Zhang, T.L.; Meng, H.; Zhou, X.M.; Jiang, C.B. A study on laminated structures in Terfenol-D/Epoxy particulate composite with enhanced magnetostriction. *J. Appl. Phys.* **2014**, *115*, 243909. [CrossRef]
90. Zhou, Z.G.; Li, J.H.; Bao, X.Q.; Liu, M.; Gao, X.X. Improvement of mechanical properties of magnetostrictive Tb-Dy-Fe alloys via preparing sintered material with low-melting Dy-Cu alloy binder. *J. Alloys Compd.* **2022**, *895*, 162572. [CrossRef]
91. Arout Chelvane, J.; Sherly, A.; Palit, M.; Talapatra, A.; Mohanty, J. Magnetic anisotropy and magnetostrictive properties of sputtered Tb-Dy-Fe-Co thin films. *J. Mater. Sci. Mater. Electron.* **2019**, *30*, 8989–8995. [CrossRef]
92. Panduranga, M.K.; Lee, T.; Chavez, A.; Prikhodko, S.V.; Carman, G.P. Polycrystalline Terfenol-D thin films grown at CMOS compatible temperature. *AIP Adv.* **2018**, *8*, 056404. [CrossRef]
93. Lee, C.H.; Chang, W.C.; Anbalagan, A.K. Anomalous X-ray scattering study on oxidized $Tb_xDy_{1-x}Fe_{2-y}$ thin films: Influence of thermal annealing on the oxide composition. *Radiat. Phys. Chemistry* **2020**, *175*, 108915. [CrossRef]
94. Panduranga, M.K.; Xiao, Z.Y.; Schneider, J.D.; Lee, T.; Klewe, C.; Chopdekar, R.; Shafer, P.; N'Diaye, A.T.; Arenholz, E.; Candler, R.N.; et al. Single magnetic domain Terfenol-D microstructures with passivating oxide layer. *J. Magn. Magn. Mater.* **2021**, *528*, 167798. [CrossRef]
95. Shim, H.; Sakamoto, K.; Inomata, N.; Toda, M.; Toan, N.V.; Ono, T. Magnetostrictive Performance of Electrodeposited $Tb_xDy_{1-x}Fe_y$ Thin Film with Microcantilever Structures. *Micromachines* **2020**, *11*, 523. [CrossRef]
96. Yu, C.Q.; Niu, R.; Peng, Z.D.; Li, H.; Luo, Y.M.; Zhou, T.J.; Dong, C.H. A Current Sensor Based on Capillary Microresonator Filled with Terfenol-D Nanoparticles. *IEEE Photonics Technol. Lett.* **2021**, *33*, 239–242. [CrossRef]
97. Sun, Y.; Wang, W.; Jia, Y.N.; Fan, S.Y.; Liang, Y. Sensitivity Improvement of TbDyFe Thin-Film Coated Saw-Based Current Sensor. In Proceedings of the 2019 14th Symposium on Piezoelectrcity, Acoustic Waves and Device Applications (Spawda19), Shijiazhuang, China, 11–14 January 2019; pp. 519–522.
98. Shen, X.L.; Sun, H.Y.; Sang, L.W.; Imura, M.; Koide, Y.; Koizumi, S.; Liao, M.Y. Integrated TbDyFe Film on a Single-Crystal Diamond Microelectromechanical Resonator for Magnetic Sensing. *Phys. Status Solidi-Rapid Res. Lett.* **2021**, *15*, 2100352. [CrossRef]
99. Sakon, T.; Matsumoto, T.; Komori, T. Rotation angle sensing system using magnetostrictive alloy Terfenol-D and permanent magnet. *Sens. Actuators A* **2021**, *321*, 112588. [CrossRef]
100. Zhu, L.L.; Li, K.S.; Luo, Y.; Yu, D.B.; Wang, Z.L.; Wu, G.Y.; Xie, J.J.; Tang, Z.F. Magnetostrictive properties and detection efficiency of TbDyFe/FeCo composite materials for nondestructive testing. *J. Rare Earths* **2019**, *37*, 166–170. [CrossRef]
101. Rajagopal, M.C.; Sinha, S. Design and analysis of magnetostrictive sensors for wireless temperature sensing. *Rev. Sci. Instrum.* **2021**, *92*, 014901. [CrossRef]
102. Kaplan, N.; Jasenek, J.; Cervenova, J.; Usakova, M. Magnetic Optical FBG Sensors Using Optical Frequency-Domain Reflectometry. *IEEE Trans. Magn.* **2019**, *55*, 1–4. [CrossRef]
103. Kaplan, N.; Jasenek, J.; Cervenova, J. The Influence of Magnetic Field Applied on Fiber Bragg Gratings. *Aip. Conf. Proc.* **2018**, *1996*, 020024. [CrossRef]

104. Karanja, J.M.; Dai, Y.T.; Zhou, X.; Liu, B.; Yang, M.H.; Dai, J.X. Femtosecond Laser Ablated FBG Multitrenches for Magnetic Field Sensor Application. *IEEE Photonics Technol. Lett.* **2015**, *27*, 1717–1720. [CrossRef]
105. Shao, Z.H.; Qiao, X.G.; Rong, Q.Z.; Sun, A. Fiber-optic magnetic field sensor using a phase-shifted fiber Bragg grating assisted by a TbDyFe bar. *Sens. Actuators A* **2017**, *261*, 49–55. [CrossRef]
106. Feng, X.X.; Jiang, Y.; Zhang, H. A mechanical amplifier based high-finesse fiber-optic Fabry-Perot interferometric sensor for the measurement of static magnetic field. *Meas. Sci. Technol.* **2021**, *32*, 125106. [CrossRef]

Article

Researching a Moving Target Detection Method Based on Magnetic Flux Induction Technology

Chaoqun Xu [1,*], Li Yang [2,*], Kui Huang [1,*], Yang Gao [1], Shaohua Zhang [1], Yuting Gao [3], Lifei Meng [1], Qi Xiao [1], Chaobo Liu [1], Bin Wang [1] and Zhong Yi [1]

[1] Beijing Institute of Spacecraft Environment Engineering, Beijing 100094, China; gygaoyang@126.com (Y.G.); xu071025@163.com (S.Z.); liu4032@126.com (L.M.); XCW@ief.ac.cn (Q.X.); chen_jingang@hotmail.com (C.L.); wangbin1983@hotmail.com (B.W.); yizhong6808@sina.com (Z.Y.)
[2] College of Electronic and Information, Southwest Minzu University, Chengdu 610041, China
[3] Institute of Remote Sensing Satellite, CAST, Beijing 100094, China; gyting1214@163.com
* Correspondence: xucq111@163.com (C.X.); swun_yangli@163.com (L.Y.); 13661966904@163.com (K.H.); Tel.: +86-010-6874-6077 (C.X.)

Citation: Xu, C.; Yang, L.; Huang, K.; Gao, Y.; Zhang, S.; Gao, Y.; Meng, L.; Xiao, Q.; Liu, C.; Wang, B.; et al. Researching a Moving Target Detection Method Based on Magnetic Flux Induction Technology. *Metals* **2021**, *11*, 1967. https://doi.org/10.3390/met11121967

Academic Editor: Aphrodite Ktena

Received: 16 August 2021
Accepted: 14 November 2021
Published: 7 December 2021

Publisher's Note: MDPI stays neutral with regard to jurisdictional claims in published maps and institutional affiliations.

Copyright: © 2021 by the authors. Licensee MDPI, Basel, Switzerland. This article is an open access article distributed under the terms and conditions of the Creative Commons Attribution (CC BY) license (https://creativecommons.org/licenses/by/4.0/).

Abstract: The ocean is a very important arena in modern warfare where all marine powers deploy their military forces. Due to the complex environment of the ocean, underwater equipment has become a very threatening means of surprise attack in modern warfare. Therefore, the timely and effective detection of underwater moving targets is the key to obtaining warfare advantages and has important strategic significance for national security. In this paper, magnetic flux induction technology was studied with regard to the difficulty of detecting underwater concealed moving targets. Firstly, the characteristics of a magnetic target were analyzed and an equivalent magnetic dipole model was established. Secondly, the structure of the rectangular induction coil was designed according to the model, and the relationship between the target's magnetism and the detection signal was deduced. The variation curves of the magnetic flux and the electromotive force induced in the coil were calculated by using the numerical simulation method, and the effects of the different motion parameters of the magnetic dipole and the size parameters of the coil on the induced electromotive force were analyzed. Finally, combined with the wavelet threshold filter, a series of field tests were carried out using ships of different materials in shallow water in order to verify the moving target detection method based on magnetic flux induction technology. The results showed that this method has an obvious response to moving targets and can effectively capture target signals, which verifies the feasibility of the magnetic flux induction detection technology.

Keywords: moving target; magnetic dipole; magnetic flux induction; induced electromotive force; numerical simulation; field test in shallow water

1. Introduction

The ocean is a very important stage in modern warfare where all countries deploy various military forces [1,2]. With the development of science and technology, modern marine warfare not only involves the competition of various advanced technologies, such as electronic and information technologies, but has also evolved into multiple space dimensions, including confrontations under water, on the sea's surface, in the air, and even in space [3]. Among these, various types of underwater moving targets play important roles. Underwater moving targets, which are shielded by the vast ocean, are the most concealed equipment in the modern naval equipment system. Since the Second World War, the research into and development of technology for detecting underwater moving targets has progressed rapidly, especially regarding the application of nuclear technology, which makes underwater equipment a very threatening means of surprise attack in modern warfare. Various countries around the world, especially maritime powers, are currently developing and deploying different types of underwater targets. In modern marine warfare, the key to obtaining a warfare advantage is to find underwater moving targets in a timely and effective manner.

In the development of underwater moving target detection technology, optical, electrical, thermal, and other technical means have been applied [4–12]. Among these, the acoustic signal detection of underwater man-made objects has become the most widely used detection method. Sonar technology [13–15] was the first mature technology to be applied to underwater detection. It uses the propagation characteristics of sound waves in water to complete the task of detecting large underwater targets through electroacoustic conversion and signal processing. However, sonar detection has inherent disadvantages. It is easy for active sonar to expose its own position to the enemy, which has become its fatal disadvantage. Passive sonar detects the target only when it emits a relatively large noise. Due to the complex marine environment, the accuracy of the target resolution of sonar detection is limited. In addition, sonar detection has its own limitations, such as "sound shadow area". With the rapid development and application of stealth technology, low noise and high acoustic stealth have evolved into revolutionary metrics in underwater moving target design. According to the data, the noise caused by the new underwater moving targets that have been continuously launched by the United States, Russia, and other countries in recent years has been lower than that of the marine background environment, meaning that it is difficult to find, identify, and track them by sonar. Therefore, there is an urgent need for new underwater moving target detection methods to make up for the shortcomings of the existing detection technologies.

Based on the principle of magnetic flux induction technology, this paper proposes a passive method that can be used to obtain and analyze the change in magnetic flux signal caused by underwater moving targets. According to the characteristics of moving targets, a magnetic dipole model [16–19] is established, the expressions of the magnetic flux of moving targets passing through a rectangular coil and the change in the electromotive force induced in the coil are deduced and calculated, and the influence of the various characteristic parameters on the induced electromotive force is analyzed. Finally, characteristic parameters such as the speed, depth, and magnetic moment of the targets can be obtained. This method has the advantages of a short execution time, lower impact from the complex shallow sea environment, all-weather working ability, and low cost. It is of great significance for providing early warnings of moving targets in coastal water areas, for island defense, and for providing early warnings in key areas.

2. Methods

When a moving target is close to a detection coil, the magnetic field of this target can be simulated by multiple magnetic dipoles; when the distance is far—that is, when the distance is greater than 3 times the size of the target—it can be treated as one magnetic dipole. This paper mainly focuses on the long-distance situation, so the magnetic field of a moving target can be simplified into one magnetic dipole model. The following is our research on the response relationship between the detection signal and the characteristics of the moving target based on this model.

2.1. Magnetic Flux Density of the Magnetic Dipole at Any Point in Space

In order to quantitatively analyze the change in electromotive force induced in a magnetic moving target in a detection closed coil, the magnetic dipole was simplified as a circular current and its coordinate system was established. As shown in Figure 1, the origin O of the coordinate system is located in the center of the magnetic dipole, the z-axis points in the direction of the magnetic moment vector, and the x-axis and y-axis point in accordance with the right-hand rule. E is a point in space whose spherical coordinate is $E(r, \varphi_0, \theta_0)$.

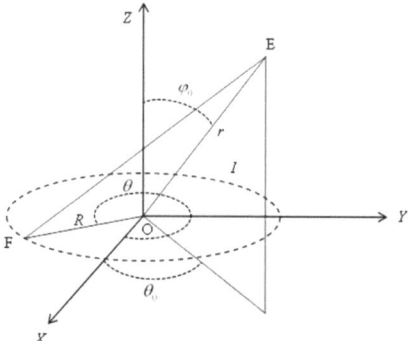

Figure 1. The coordinate system and the magnetic dipole.

Let the intensity of the circular current be I with unit A. The radius of a circle is R with unit m and r is the distance from the center O to a point in space with unit m. According to the Biot–Savart law [20] and the concept of magnetic moment [21], the following equations can be obtained:

$$\begin{cases} B_x = \dfrac{3\mu}{8\pi} \dfrac{p_m}{\left(\sqrt{R^2+r^2}\right)^3} \dfrac{r^2 \sin 2\varphi_0 \cos\theta_0}{(R^2+r^2)} \\ B_y = \dfrac{3\mu}{8\pi} \dfrac{p_m}{\left(\sqrt{R^2+r^2}\right)^3} \dfrac{r^2 \sin 2\varphi_0 \sin\theta_0}{(R^2+r^2)} \\ B_z = \dfrac{\mu}{2\pi} \dfrac{p_m}{\left(\sqrt{R^2+r^2}\right)^3} (1 - \dfrac{3}{2} \dfrac{r^2 \sin^2\varphi_0}{R^2+r^2}) \end{cases} \quad (1)$$

This is a set of expressions of the magnetic flux density of the space point $E(r, \varphi_0, \theta_0)$ in three directions. For the far field—i.e., $R \ll r$—expressions can be simplified by eliminating R. The x and y components of the magnetic flux density at any point in space are related to p_m, r, φ_0, and θ_0, while the z component is only related to p_m, r, φ_0, and θ_0.

2.2. Expression Derivation of Magnetic Flux in the Rectangular Coil at a Certain Time

For a rectangular detection coil with length a and width b whose number of turns is N, a rectangular coordinate system is established with the center of the coil as the zero point, as shown in Figure 2. The point $Q(x_q, 0, z_q)$ in the figure is the moving target with the magnetic moment m. The target Q passes along the positive direction of the x-axis at a uniform speed v directly above the coil. $P(x_P, y_P, 0)$ represents any point within the rectangular coil. Since this paper focuses on the long-distance situation—that is, $R \ll r$—the magnetic flux density of the magnetic dipole Q at point P can be calculated using Equation (1). Since the magnetic flux passing through the rectangular coil is only related to the magnetic flux density perpendicular to the plane where the rectangular coil is located, it can be divided into three cases according to the different magnetic moment directions of the magnetic dipole.

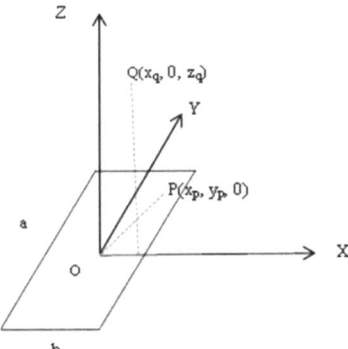

Figure 2. The coordinate system and the rectangular detection coil.

2.2.1. The Direction of Magnetic Moment in Positive X Direction

According to the coordinate transformation relationship, the magnetic flux in the rectangular coil was related to B_x in the magnetic dipole coordinate system shown in Figure 2, where $R = 0$, $r = \sqrt{(x_p - x_q)^2 + y_p^2 + z_q^2}$. The following expression could be obtained by substituting R and r into Equation (1):

$$B_x = \frac{3\mu m}{8\pi} \frac{1}{\left[\sqrt{(x_p - x_q)^2 + y_p^2 + z_q^2}\right]^3} \sin 2\varphi \cos\theta \qquad (2)$$

where:

$$\sin\varphi = \frac{\sqrt{y_p^2 + z_q^2}}{\sqrt{(x_p - x_q)^2 + y_p^2 + z_q^2}}$$

$$\sin\varphi = \frac{\sqrt{y_p^2 + z_q^2}}{\sqrt{(x_p - x_q)^2 + y_p^2 + z_q^2}}$$

$$\cos\theta = \frac{z_q}{\sqrt{y_p^2 + z_q^2}}$$

The magnetic flux through the rectangular coil can be obtained as follows:

$$\Phi = \int_{-\frac{a}{2}}^{\frac{a}{2}} \int_{-\frac{b}{2}}^{\frac{b}{2}} B_z dx dy = -\frac{3\mu m}{4\pi} \int_{-\frac{a}{2}}^{\frac{a}{2}} \int_{-\frac{b}{2}}^{\frac{b}{2}} \left[(x - x_q)^2 + y^2 + z_q^2\right]^{-\frac{5}{2}} \cdot (x - x_q) \cdot z_q dx dy \qquad (3)$$

The above equation shows the magnetic flux generated in the rectangular coil by the magnetic dipole whose magnetic moment is in the positive x direction.

2.2.2. The Magnetic Moment in Positive Y Direction

According to the coordinate transformation relationship, the magnetic flux in the rectangular coil is related to B_y in the magnetic dipole coordinate system shown in Figure 2:

$$B_y = \frac{3\mu m}{8\pi} \frac{1}{\left[\sqrt{(x_p - x_q)^2 + y_p^2 + z_q^2}\right]^3} \sin 2\varphi \sin\theta \qquad (4)$$

where:

$$\sin\varphi = \frac{\sqrt{(x_p - x_q)^2 + z_q^2}}{\sqrt{(x_p - x_q)^2 + y_p^2 + z_q^2}}$$

$$\cos\varphi = \frac{y_p}{\sqrt{(x_p - x_q)^2 + y_p^2 + z_q^2}}$$

$$\sin\theta = \frac{-z_q}{\sqrt{(x_p - x_q)^2 + z_q^2}}$$

The magnetic flux through the rectangular coil can be obtained as follows:

$$\Phi = \int_{-\frac{a}{2}}^{\frac{a}{2}} \int_{-\frac{b}{2}}^{\frac{b}{2}} B_z dxdy = -\frac{3\mu m}{4\pi} \int_{-\frac{a}{2}}^{\frac{a}{2}} \int_{-\frac{b}{2}}^{\frac{b}{2}} B_y dxdy = 0 \tag{5}$$

It can be seen that when the magnetic moment of the magnetic dipole is in the positive y direction, the magnetic flux generated in the rectangular coil is zero.

2.2.3. The Magnetic Moment in Positive Z Direction

According to the coordinate transformation relationship, the magnetic flux in the rectangular coil is related to B_z in the magnetic dipole coordinate system shown in Figure 2:

$$B_z = \frac{\mu m}{2\pi} \frac{1}{\left[\sqrt{(x_p - x_q)^2 + y_p^2 + z_q^2}\right]^3} (1 - \frac{3}{2}\sin^2\varphi) \tag{6}$$

where:

$$\sin\varphi = \frac{\sqrt{(x_p - x_q)^2 + y_p^2}}{\sqrt{(x_p - x_q)^2 + y_p^2 + z_q^2}}$$

The magnetic flux through the rectangular coil can be calculated using the following equation:

$$\Phi = \int_{-\frac{a}{2}}^{\frac{a}{2}} \int_{-\frac{b}{2}}^{\frac{b}{2}} B_z dxdy = \int_{-\frac{a}{2}}^{\frac{a}{2}} \int_{-\frac{b}{2}}^{\frac{b}{2}} \frac{\mu m}{2\pi} \frac{1}{\left[\sqrt{(x-x_q)^2 + y^2 + z_q^2}\right]^3} (1 - \frac{3}{2}(\frac{\sqrt{(x-x_q)^2 + y^2}}{\sqrt{(x-x_q)^2 + y^2 + z_q^2}})^2) dxdy$$

$$= \frac{\mu m}{4\pi} \int_{-\frac{a}{2}}^{\frac{a}{2}} \int_{-\frac{b}{2}}^{\frac{b}{2}} \left[(x-x_q)^2 + y^2 + z_q^2\right]^{-\frac{5}{2}} \cdot \left[-(x-x_q)^2 - y^2 + 2z_q^2\right] dxdy \tag{7}$$

The above equation is the magnetic flux generated in the rectangular coil by the magnetic dipole whose magnetic moment is in the positive z direction.

According to Faraday's law of electromagnetic induction, the induced electromotive force at time t can be obtained by differentiating the magnetic flux $\varepsilon = n\frac{\Delta\phi}{\Delta t}$ with respect to time t, where n is the number of coil turns.

3. Simulation Calculation

3.1. Influence of the Target's Magnetic Moment Direction on Induced Electromotive Force

Suppose a magnetic dipole with a magnetic moment of 50 A·m^2 whose height from the coil is h = 20 m is initially located at -100 m. This magnetic dipole moves along the x-axis with a dynamic speed of v = 1 m/s for 200 s. The coil has a length of a = 20 m and a width of b = 5 m, with a number of turns N = 100 and a magnetic permeability of $\mu = 4\pi \times 10^{-7}$ H/m.

When the magnetic moment of the magnetic dipole is in the positive x direction, the simulation results are as shown in the figures below.

It can be seen from Figure 3 that for the magnetic moment of the magnetic dipole pointing in the positive x direction, the magnetic flux generated in the rectangular coil first increases from zero and then decreases back to zero when the magnetic dipole is right below the coil at 100 s. After that, it again increases and then decreases to zero in the opposite direction. The induced electromotive force first increases in the opposite direction from zero to a negative value, then increases in the positive direction, reaching the maximum when the dipole is right below the coil at 100 s. After that, it decreases back to a negative value and finally returns to zero.

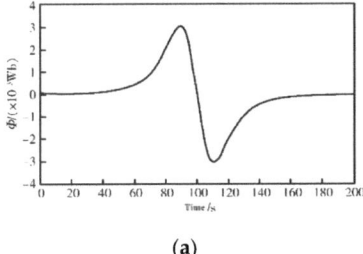 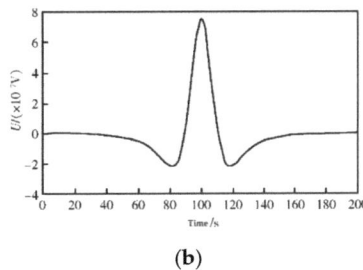

(a) (b)

Figure 3. (a) The change of magnetic flux along the x-axis; (b) the change in induced electromotive force along the x-axis.

When the magnetic moment of the magnetic dipole is in the positive z direction, the simulation results are as shown in the figures below.

It can be seen from Figure 4 that for the magnetic moment of the magnetic dipole pointing in the positive z direction, the magnetic flux generated in the rectangular coil first increases a small amount from zero in the positive direction, then increases continuously in the opposite direction, reaching the maximum when the magnetic dipole is right below the coil at 100 s. After that, it decreases to a certain positive value and finally returns to zero. The induced electromotive force first increases from zero and then decreases back to zero when the magnetic dipole is right below the coil at 100 s. After that, it again increases and then decreases to zero in the opposite direction. For a magnetic dipole whose magnetic moment is in an arbitrary direction, when calculating the induced electromotive force its magnetic moment should be decomposed along the coordinate axes. Calculations should be carried out separately and the overall induced electromotive force should be superimposed.

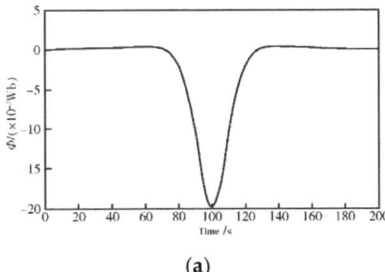 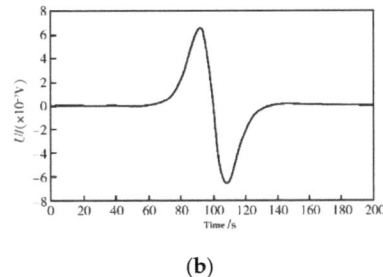

(a) (b)

Figure 4. (a) The change law of magnetic flux of the z-axis; (b) the change law of induced electromotive force of the z-axis.

3.2. Influence of the Target's Motion Parameters on Induced Electromotive Force

In this section, two parameters, the height h of the magnetic dipole from the coil and its moving speed v, were selected in order to analyze their influence on the induced electromotive force. For simplification, the magnetic moment of the magnetic dipole was set to pointing in the positive x direction.

In four different simulation cases, the height was set to $z_q = 20$ m, $z_q = 25$ m, $z_q = 30$ m, and $z_q = 35$ m, respectively, and other parameters were kept the same as those in Section 3.1.

Figure 5 shows the influence of the magnetic dipole's height on the induced electromotive force. It can be seen that the induced absolute value of the electromotive force decreases with the increase in height h. There is a power exponential relationship between the peak value of the induced electromotive force and the height, and the index is related to the coil size.

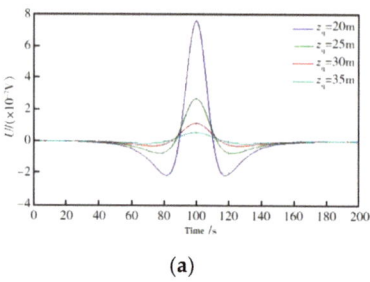
(a)

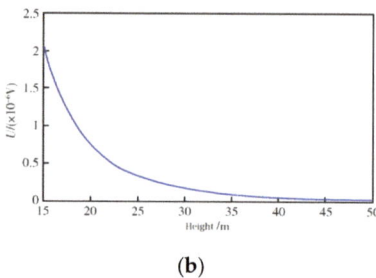
(b)

Figure 5. (a) The change curve of induced electromotive force at different heights; (b) the change in the curve peak at different heights.

Then, in four other simulation cases, the speed is set to $v = 2$ m/s, $v = 1.5$ m/s, $v = 1$ m/s, $v = 0.5$ m/s, and $v = 0.25$ m/s for the simulation calculations, and the other parameters arere the same as those in Section 3.1.

Figure 6 shows the influence of velocity on the induced electromotive force. It can be seen that the absolute value of the induced electromotive force increases with the increase in speed v. There is a linear relationship between the peak value of the induced electromotive force and the speed.

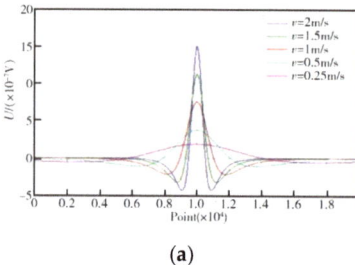
(a)

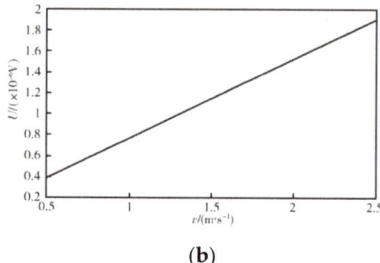
(b)

Figure 6. (a) The change curve of induced electromotive force at different speeds; (b) the change of curve peak at different speeds.

3.3. Influence of the Coil's Size Parameters on Induced Electromotive Force

According to our theoretical analysis, when the coil size and the height h from the magnetic dipole to the coil were of the same order of magnitude, the larger the coil size, the greater the induced electromotive force. Due to the complex influence situation, only two cases with coil sizes of $a = 20$ m, $b = 5$ m, and $a = 20$ m, $b = 30$ m were subjected to a comparative analysis. Other parameters were the same as those in Section 3.1.

As can be seen from Figure 7, in a certain range, when the coil size was increased, the induced electromotive force also increased to a certain extent. Through further analysis, it could be concluded that the coil size would affect the power exponential relationship between the peak value of the induced electromotive force and the magnetic dipole height h. The larger the coil size, the smaller the exponential index.

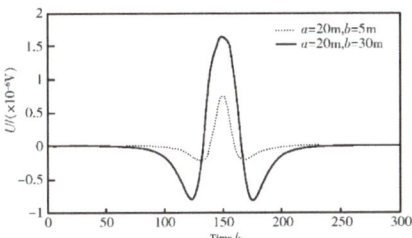

Figure 7. The influence of coil size on induced electromotive force.

4. Test Verification

After a field investigation, the research group selected a coastal area to carry out a series of field tests on the sea. In the test process, targets with different magnetic characteristics passed through the detection coil many times, and the obtained target magnetic flux signal was processed and analyzed by the wavelet threshold filtering method [22–25] to extract the required information.

4.1. The Detection Coil

The detection coil and data acquisition equipment used in the trials are shown in Figure 8. The coil wound by copper wire had a thickness of 10 cm, a frame size of 100 cm × 150 cm, and 500 turns. The copper wire diameter was 0.7 mm and the length of the conductive cable was about 150 m. An eight-channel data acquisition system with 128 Gb of memory was adopted. The coil was put into the sea at a depth of 10 m. Various types of ships passed over the coil many times at different speeds and from different distances so as to obtain the target flux signals under different motion states.

Figure 8. Detection coil and data acquisition equipment.

4.2. Result Analysis of the Extracted Magnetic Flux Signals of the Wooden Ship

As shown in Figure 9, a wooden ship passed over the detection coil in the field tests.

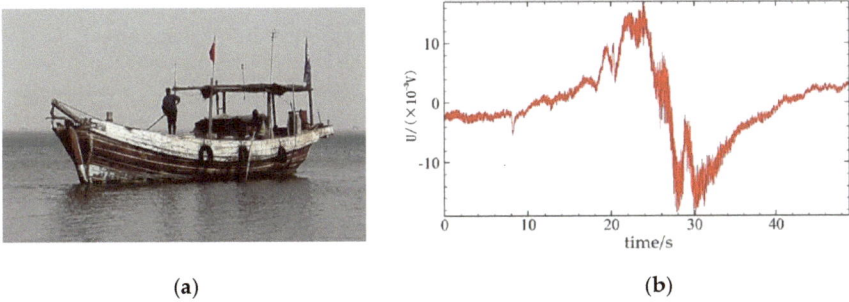

(a) (b)

Figure 9. (a) The wooden ship test; (b) magnetic flux signal of the wooden ship.

It can be seen from the figure that the curve of the target magnetic flux signal fluctuated obviously, but some local features were covered by the background noise. The wavelet threshold filtering method was used to process the magnetic flux signal. The low-frequency part and the high-frequency part after thresholding were reconstructed separately. The results are shown in Figure 10.

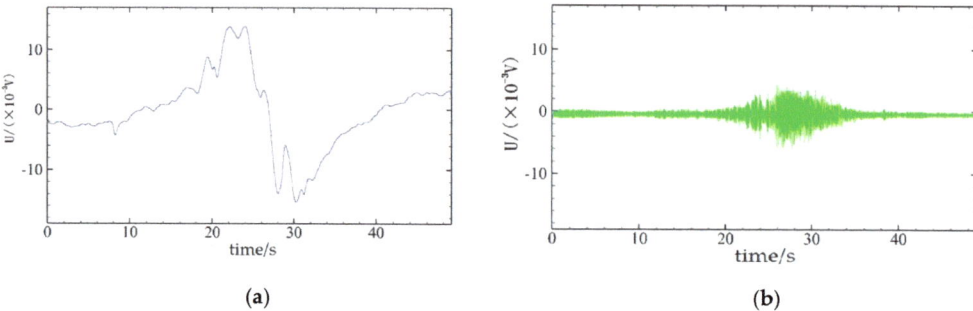

Figure 10. (**a**) Low-frequency part of the magnetic flux signal of the wooden ship; (**b**) high-frequency part of the magnetic flux signal of the wooden ship.

After analysis, it could be concluded that the low-frequency part of the magnetic flux signal was induced by the wooden ship's bottom magnetic field. The magnetic characteristics of the reconstructed high-frequency part of the original signal had an obvious periodicity, which was related to the rotation of the wooden ship's engine.

4.3. Result Analysis of the Extracted Magnetic Flux Signal of the Speedboat

As shown in Figure 11, a speedboat passed over the detection coil in the field tests.

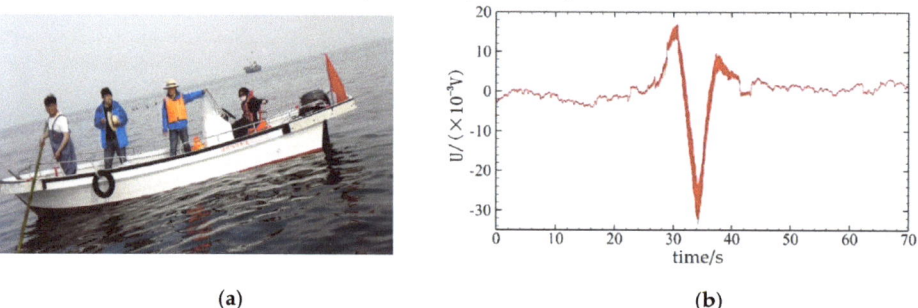

Figure 11. (**a**) The speedboat test; (**b**) magnetic flux signal of the speedboat.

It can be seen from the figure that the curve of the target magnetic flux signal fluctuated obviously, but some local features were covered by the background noise. The wavelet threshold filtering method was used to process the magnetic flux signal. The low-frequency part and the high-frequency part after thresholding were reconstructed separately. The results are shown in Figure 12.

After analysis, it could be concluded that the low-frequency part of the magnetic flux signal was induced by the speedboat's bottom magnetic field. The signal curve of the speedboat was quite different from the wooden ship. It was easy to distinguish these two kinds of targets. The magnetic characteristics of the reconstructed high-frequency part of the original signal had an obvious periodicity, which was related to the rotation of the speedboat's engine.

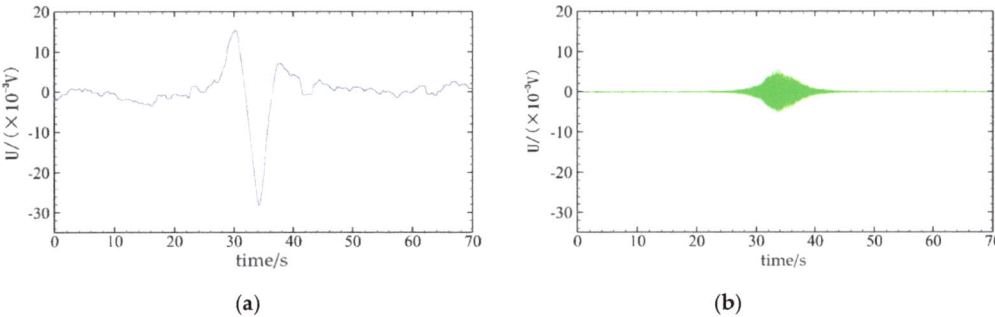

Figure 12. (**a**) Low-frequency part of the magnetic flux signal of the speedboat; (**b**) high-frequency part of the magnetic flux signal of the speedboat.

4.4. Result Analysis of the Extracted Magnetic Flux Signal of the Rubber Boat

As shown in Figure 13, a rubber boat was used to pass over the detection coil.

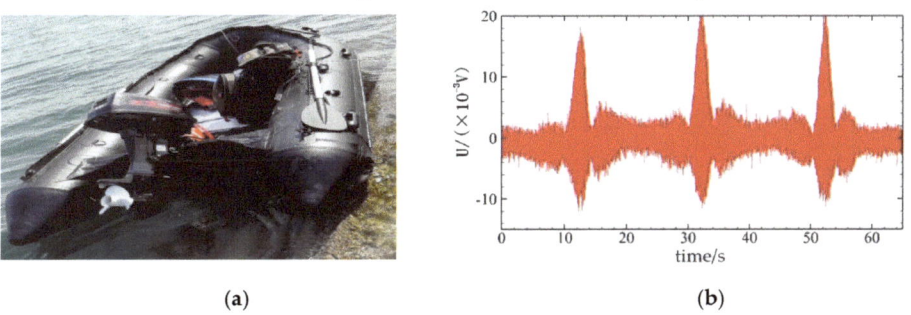

Figure 13. (**a**) The rubber boat test; (**b**) magnetic flux signal of the rubber boat.

It can be seen from the figure that the curve of the target magnetic flux signal fluctuated obviously, but some local features were covered by the background noise. The wavelet threshold filtering method was used to process the magnetic flux signal. The low-frequency part and the high-frequency part after thresholding were reconstructed separately. The results are shown in Figure 14.

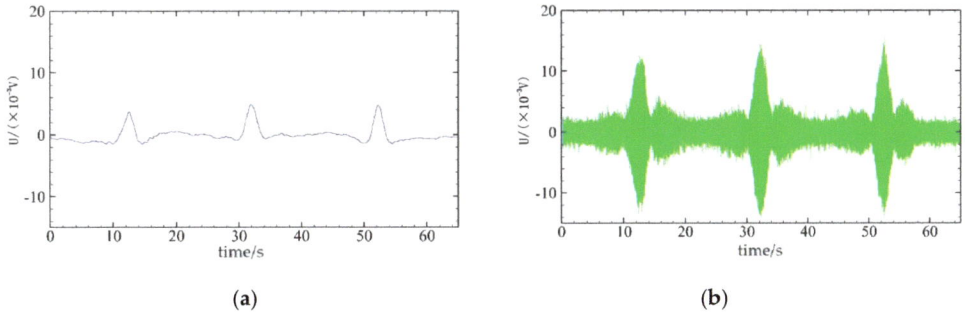

Figure 14. (**a**) Low-frequency part of the magnetic flux signal of the rubber boat; (**b**) high-frequency part of the magnetic flux signal of the rubber boat.

After analysis, it could be concluded that the low-frequency part of the magnetic flux signal was induced by the rubber boat's bottom magnetic field. The signal curve of the rubber boat was quite different from that of the other targets. It was easy to distinguish these three kinds of targets. The magnetic characteristics of the reconstructed high-frequency part of the original signal had an obvious periodicity, which was related to the rotation of the rubber boat's engine.

5. Conclusions

This paper established a magnetic dipole model and designed coil structure sizes according to the characteristics of hidden moving targets. The magnetic flux and electromotive force induced in the rectangular coil by the magnetic dipole were deduced in detail, and their variation curves were obtained using the numerical simulation method. The influence of the height h, the velocity V, and the detection coil size parameters on the induced electromotive force were analyzed. The conclusions were as follows:

(1) The induced electromotive force increases with the decrease in the target's height. There is a power exponential relationship between the peak value of the induced electromotive force and the height.
(2) The induced electromotive force increases with the increase in the target's velocity. There is a linear relationship between the peak value of the induced electromotive force and the velocity.
(3) The induced electromotive force increases with the increase in the detection coil's size within a certain range.

Finally, in order to verify the feasibility of the magnetic flux induction detection technology, a series of field tests with ship targets of different materials were carried out in the sea and the wavelet threshold filtering method was used in the test data analysis, which provided guidance for moving target detection and coil design optimization in the future. In addition, due to the complexity and variability of the characteristics of the moving targets and the marine environment, this subject still needs further research regarding the coil size, moving target characteristics, and data inversion.

Author Contributions: Conceptualization, Z.Y. and L.Y.; methodology, K.H. and Y.G. (Yuting Gao); software, B.W. and S.Z.; validation, C.L. and C.X.; formal analysis, Q.X.; investigation, C.X. and Y.G. (Yang Gao); resources, K.H.; data curation, K.H. and Y.G. (Yuting Gao); writing—original draft preparation, C.X.; writing—review and editing, C.X. and Y.G. (Yuting Gao); visualization, K.H.; supervision, Z.Y.; project administration, L.M. All authors have read and agreed to the published version of the manuscript.

Funding: This research received no external funding.

Institutional Review Board Statement: Not applicable.

Informed Consent Statement: Not applicable.

Data Availability Statement: Not applicable.

Conflicts of Interest: The authors declare no conflict of interest.

References

1. Kohler, M.M. The Joint Force Maritime Component Command and the Marine Corps: Integrate to Win the Black Sea Fight. *J. Adv. Mil. Stud.* **2021**, *11*, 88–105. [CrossRef]
2. Li, J.C.; Zhou, Z.C. An Analysis of the Decline of British Overseas Military Bases. *J. Liaoning Univ. (Philos. Soc. Sci.)* **2020**, *48*, 152–158.
3. Fan, X.Y.; He, X.Y.; Yang, L. Research on application of ocean remote sensing in military marine environment guarantee. *J. Nav. Univ. Eng.* **2020**, *17*, 39–42.
4. Diamant, R.; Kipnis, D.; Bigal, E. An Active Acoustic Track-Before-Detect Approach for Finding Underwater Mobile Targets. *IEEE J. Sel. Top. Signal Process.* **2019**, *13*, 104–119. [CrossRef]
5. Wang, X.M.; Liu, Z.P.; Sun, J.C. Sonar Image Detection Algorithm Based on Two-Phase Manifold Partner Clustering. *J. Harbin Inst. Technol.* **2015**, *4*, 105–114.
6. Venturino, L.; Grossi, E.; Lops, M. Track-Before-Detect for Multiframe Detection with Censored Observations. *IEEE Trans. Aerosp. Electron. Syst.* **2014**, *50*, 2032–2046.

7. Zhu, J.J.; Yu, S.; Gao, L. Saliency-Based Diver Target Detection and Localization Method. *Math. Probl. Eng.* **2020**, *2020*, 3186834. [CrossRef]
8. Zhao, Y.L.; Yan, P.; Li, X.; Tan, B.; Wei, P. Research on the equivalent relationship of torpedo penetrated by underwater supercavitation projectile based on energy consumption model. *J. Phys. Conf. Ser.* **2020**, *1507*, 032059. [CrossRef]
9. Jawahar, A. More Reliable and Automated Target Localisation when Tracking Low Signature Targets in Areas with Heavy Shipping. *Artif. Intell. Syst. Mach. Learn.* **2016**, *8*, 299–303.
10. Egozi, A. Israeli underwater capabilities at the cutting edge of technology. *Asia-Pac. Def. Rep.* **2018**, *44*, 42–43.
11. Chalcogens. Underwater defense: New ways to protect divers in the deep. *NewsRx Health Sci.* **2017**, 283–285.
12. Jawahar, A. Target Localization and Mathematical Modelling for Maritime Based Sonar Applications Using Kalman Filter. *Autom. Auton. Syst.* **2016**, *8*, 287–298.
13. Udovydchenkov, I.A.; Stephen, R.A.; Howe, B.M. Bottom interacting sound at 50 km range in a deep ocean environment. *J. Acoust. Soc. Am.* **2012**, *132*, 2224–2231. [CrossRef]
14. Moore, P.W.; Lane, D.M.; Capus, C. Bio-inspired wideband sonar signals based on observations of the bottlenose dolphin (*Tursiops truncatus*). *J. Acoust. Soc. Am.* **2007**, *121*, 594–604.
15. Ferla, C.; Porter, M.B. Receiver depth selection for passive sonar systems. *IEEE J. Ocean. Eng. A J. Devoted Appl. Electr. Electron. Eng. Ocean. Environ.* **1991**, *16*, 267–278. [CrossRef]
16. Salem, A.; Ushijima, K. Automatic detection of UXO from airborne magnetic data using a neural network. *Subsurf. Sens. Technol. Appfications* **2001**, *2*, 191–213. [CrossRef]
17. Chen, J.J.; Yi, Z.; Meng, L.F. Multi-dipole discrimination technology based on Euler inverse method. *Spacecr. Environ. Eng.* **2013**, *30*, 401–406.
18. Tang, J.F.; Gong, S.G.; Wang, J.G. Target Positioning and Parameter Estimation Based on Magnetic Dipole Model. *Acta Electron. Sin.* **2002**, *30*, 614–616.
19. Zhang, C.Y.; Xiao, C.H.; Gao, J.J. Experiment Research of Magnetic Dipole Model Applicability for a Magnetic Object. *J. Basic Sci. Eng.* **2010**, *18*, 862–868.
20. Takayuki, I.; Akihiro, S.; Masaharu, K. Magnetic dipole signal detection and location using subspace method. *Electron. Commun. Jpn.* **2002**, *85*, 23–34.
21. Ren, L.P.; Zhao, J.S.; Hou, S.X. The Magnetic Field Space Distribution Pattern of Magnetic Diples. *Hydrogr. Surv. Charting* **2002**, *22*, 18–21.
22. Hsung, T.C.; Lun, D.P.K.; Siu, W.C. Denoising by singularity detection. *IEEE Trans. Signal Proc.* **1999**, *47*, 3139–3144. [CrossRef]
23. Xu, Y.S.; Weaver, J.B.; Healy, D.M. Wavelet transform domain filters: A spatially selective noise filtration technique. *IEEE Trans. Image Proc.* **1994**, *3*, 747–758.
24. Mallat, S. A theory for multiresolution signal decomposition: The wavelet representation. *IEEE Trans. Pattern Anal Mach. Intel.* **1989**, *11*, 674–692. [CrossRef]
25. Rioul, O.; Vetterli, M. Wavelets and signal processing. *IEEE Signal Process. Mag.* **1991**, *8*, 14–38. [CrossRef]

Article

Correction Method of Three-Axis Magnetic Sensor Based on DA–LM

Li Yang [1,*,†], Caihong Li [1,†], Song Zhang [1], Chaoqun Xu [2], Hun Chen [1], Shuting Xiao [1], Xiaoyu Tang [1] and Yongxin Li [1]

1. College of Electrical & Information Engineering, Southwest Minzu University, Chengdu 610041, China; licaihong@stu.swun.edu.cn (C.L.); song@stu.swun.edu.cn (S.Z.); chenhui@swun.cn (H.C.); xiaoshuting@stu.swun.edu.cn (S.X.); tangxiaoyu@stu.swun.edu.cn (X.T.); liyongxin@stu.swun.edu.cn (Y.L.)
2. Beijing Institute of Spacecraft Environment Engineering, Beijing 100094, China; xucq111@163.com
* Correspondence: yangli@swun.cn
† These authors contributed equally to this work.

Citation: Yang, L.; Li, C.; Zhang, S.; Xu, C.; Chen, H.; Xiao, S.; Tang, X.; Li, Y. Correction Method of Three-Axis Magnetic Sensor Based on DA–LM. *Metals* **2022**, *12*, 428. https://doi.org/10.3390/met12030428

Academic Editor: Aphrodite Ktena

Received: 17 January 2022
Accepted: 23 February 2022
Published: 28 February 2022

Publisher's Note: MDPI stays neutral with regard to jurisdictional claims in published maps and institutional affiliations.

Copyright: © 2022 by the authors. Licensee MDPI, Basel, Switzerland. This article is an open access article distributed under the terms and conditions of the Creative Commons Attribution (CC BY) license (https:// creativecommons.org/licenses/by/ 4.0/).

Abstract: The fluxgate magnetometer has the advantages of having a small volume and low power consumption and being light weight and is commonly used to detect weak magnetic targets, including ferrous metals, unexploded bombs (UXOs), and underground corrosion pipelines. However, the detection accuracy of the fluxgate magnetometer is affected by its own error. To obtain more accurate detection data, the sensor must be error-corrected before application. Previous researchers easily fell into the local minimum when solving error parameters. In this paper, the error correction method was proposed to tackle the problem, which combines the Dragonfly algorithm (DA) and the Levenberg–Marquardt (LM) algorithm, thereby solving the problem of the LM algorithm and improving the accuracy of solving error parameters. Firstly, we analyzed the error sources of the three-axis magnetic sensor and established the error model. Then, the error parameters were solved by using the LM algorithm and DA–LM algorithm, respectively. In addition, by comparing the results of the two methods, we found that the error parameters solved by using the DA–LM algorithm were more accurate. Finally, the magnetic measurement data were corrected. The simulation results show that the DA–LM algorithm can accurately solve the error parameters of the triaxial magnetic sensor, proving the effectiveness of the proposed algorithm. The experimental results show that the difference between the corrected and the ideal total value was decreased from 300 nT to 5 nT, which further verified the effectiveness of the DA–LM algorithm.

Keywords: three-axis magnetic sensor; weak magnetic target; DA–LM algorithm; parameter estimation; error correction

1. Introduction

The Earth's magnetic field is its inherent physical field. Although it cannot be seen or felt, it is always there and closely related to human life [1]. Magnetic objects or ferromagnetic materials magnetized by the Earth's magnetic field and moving conductors cutting through the Earth's magnetic field generate the eddy current magnetic field and cause disturbances to the Earth's magnetic field, which is called magnetic anomalies. Magnetic anomaly detection can be used to detect and locate magnetic targets based on magnetic anomaly, which is a passive detection method based on basic physical phenomena. It is widely used in military antisubmarine [2], geological prospecting [3], unexploded ordnance detection [4], underwater target detection [5], space magnetic field detection [6], and medical endoscopic positioning [7] due to its excellent stability and versatility, which has extremely high military significance and civilian value. A fluxgate magnetometer is the most widely-used magnetic sensor at present, which can be used to measure a slowly moving magnetic field with good robustness and high resolution [8,9]. The three-axis fluxgate sensor can be used to obtain the total and the component information of the magnetic field simultaneously.

Due to the limitation of processing and installation craft level, the three axes of the sensor are not strictly orthogonal; the sensitivity of each axis is not exactly the same, and they all have zero drift. Therefore, the measured values of the total and component magnetic fields of the sensor are greatly different from the actual values. To obtain high-precision measurements, it is necessary to correct the sensor before its application.

There are two common error correction methods for magnetic sensors, that is, the vector correction and the scalar correction. The former is first to compare with the known magnetic field vector [10,11] and then makes correction to the measured data. However, it is difficult to obtain high-precision magnetic field vector in practical applications. The scalar correction does not require a known magnetic field vector, which is performed with a fixed value of the total magnetic field in a constant magnetic field as a constraint condition. The scalar correction method has attracted the attention of many scholars due to its advantage of easy operation in the actual environment. In [12,13], the least square method was proposed to estimate the parameters that were then brought into the error parameter model for correction. It was found that the error of the measured data was significantly suppressed. In [14], the method of leastsquares combined with winding the character "8" was proposed for rapid correction, which has the advantages of a small amount of data required, simple correction process, and good correction effect. In [15], the use of a genetic algorithm was proposed to solve the parameters in the error model, and good results were achieved.

Although the above algorithms were able to achieve good correction effects, the least square method for solving parameters in [12,13] had a great impact on the parameter compensation of abnormal points in the sensor sampling process. The angle coverage of the method of least squares combined with "8", as shown in literature [14], must be above 1.3 to obtain more accurate correction parameters. The parameters in the error model solved by using genetic algorithm in [15] are only in the simulation stage at present, and the setting of the geomagnetic field range value of the algorithm has a great influence on parameter estimation. Aiming at the shortcomings of the above error correction algorithms, this paper proposes a correction method combining the DA algorithm and the LM algorithm. The Dragonfly algorithm (DA) is a kind of bionics algorithm, which simulates the static and dynamic behavior of dragonflies in nature [16]. The Levenberg–Marquardt algorithm (LM) is an optimization method used to solve nonlinear least squares problems. It is insensitive to the overparameterization problems and can effectively deal with redundant parameters [17].

This paper is organized as follows. Section 1 analyzes the error sources of three-axis sensor and gives the error correction objective function. Section 2 introduces the basic principles of the DA algorithm and the LM algorithm and the steps of their combination. In Section 3, the simulation results show that the algorithm is accurate in determining the error parameters and has a good correction effect on the total magnetic field and the three components of the magnetic field. In Section 4, experimental data are used to verify the correction effect of the algorithm. Finally, conclusions are discussed in Section 5.

2. The Analysis of Error Model

Currently, the three-axis magnetic sensor is widely used due to its ability to obtain three components of the magnetic field and the total magnetic field simultaneously. However, due to the limitation of processing and installation technology, no axis of the three-axis sensor is strictly orthogonal, and the sensitivity is not exactly the same. In addition, each axis has zero drift [18]. Therefore, the error between the measured value and the actual one is large.

A three-axis magnetic sensor is composed of three probes that are perpendicular to each other, but the processing and installation technology cannot guarantee the complete orthogonality of the probes. As shown in Figure 1, $O - XYZ$ represents the ideal coordinate system of the three-axis magnetic sensor, and $O - X'Y'Z'$ stands for the actual coordinate system. To simplify the calculation, it is assumed that the Z axis recombines with the Z' axis, and $Z - O - Y$ is coplanar with $Z' - O - Y'$.

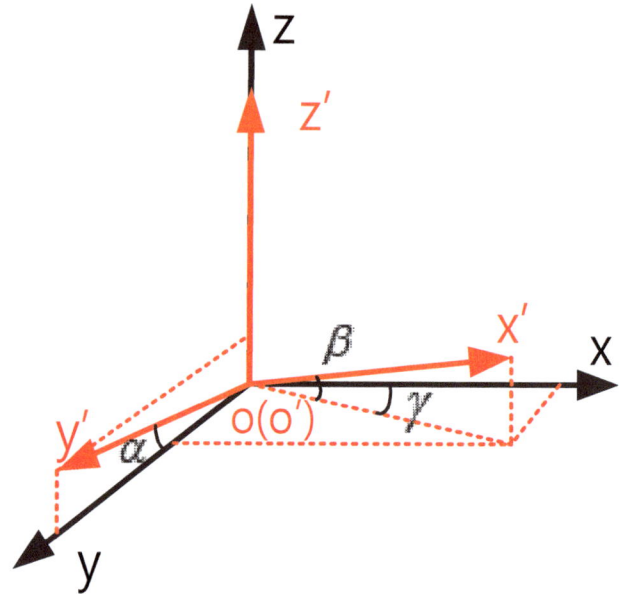

Figure 1. Nonorthogonal schematic diagram of three-axis magnetic sensor.

Where α refers to the angle between the y axis and the y' axis, β represents the angle between the x' axis and the xoy plan, and γ stands for the angle between the projection of the x' axis in the xoy plan and the x axis. If only the nonorthogonal error is considered, then

$$H_m = AH_e = \begin{bmatrix} \cos\beta\cos\gamma & \cos\beta\sin\gamma & \sin\beta \\ 0 & \cos\alpha & \sin\alpha \\ 0 & 0 & 1 \end{bmatrix} H_e \qquad (1)$$

where $H_m = \begin{bmatrix} H_{mx} \\ H_{my} \\ H_{mz} \end{bmatrix}$ refers to the measured value of the three-axis magnetic sensor,

$H_e = \begin{bmatrix} H_{ex} \\ H_{ey} \\ H_{ez} \end{bmatrix}$ represents the actual value of the three-axis magnetic sensor, and A stands

for the nonorthogonal error matrix.

The magnetic core of the three-axis magnetic sensor is made of soft magnetic material with high permeability and low coercivity, which reaches its saturation state under external excitation that is then converted to the voltage value. After that, it collects the change in the measured magnetic field. The sensor contains detector circuit, integral filter circuit, and signal feedback circuit, etc. Different characteristics of electronic components in the circuit lead to differences in the sensitivity of the three axes. If only the sensitivity factor error is considered, then

$$H_m = LH_e = \begin{bmatrix} L_x & 0 & 0 \\ 0 & L_y & 0 \\ 0 & 0 & L_z \end{bmatrix} H_e \qquad (2)$$

where L_x refers to the sensitivity factor of the x axis, L_y stands for the sensitivity factor of the y axis, L_z represents the sensitivity factor of the z axis, and L is the error matrix of the sensitivity factor.

Ideally, the sensor output should be zero in an absolutely zero magnetic field. However, the output is not zero because of the conversion zero drift of the signal processing circuit inside the sensor. If only this error is considered, then

$$H_m = H_e + h_0 \tag{3}$$

where $h_0 = \begin{bmatrix} h_{0x} \\ h_{0y} \\ h_{0z} \end{bmatrix}$ refers to zero drift.

Considering the above three errors, the error model of the three-axis magnetic sensor can be obtained as follows:

$$H_m = LAH_e + h_0 \tag{4}$$

According to Equation (4), the comprehensive error calibration model of the three-axis magnetic sensor can be obtained as follows:

$$H_e = A^{-1}L^{-1}(H_m - h_0) \tag{5}$$

Replacing the $A^{-1}L^{-1}$ by P, and $A^{-1}L^{-1}h_0$ by Q, we can obtain:

$$H_e = PH_m - Q \tag{6}$$

Assumption $P = \begin{pmatrix} m_1 & m_2 & m_3 \\ m_4 & m_5 & m_6 \\ m_7 & m_8 & m_9 \end{pmatrix}$, $Q = \begin{pmatrix} m_{10} \\ m_{11} \\ m_{12} \end{pmatrix}$, where

$$\begin{cases} m1 = \frac{1}{L_x \cdot \cos\beta \cdot \cos\gamma} \\ m2 = \frac{-\sin\gamma}{L_y \cdot \cos\alpha \cdot \cos\gamma} \\ m3 = \frac{-(\cos\alpha \cdot \sin\beta - \cos\beta \cdot \sin\alpha \cdot \sin\gamma)}{L_z \cdot \cos\alpha \cdot \cos\beta \cdot \cos\gamma} \\ m4 = 0 \\ m5 = \frac{1}{L_y \cdot \cos\alpha} \\ m6 = \frac{-\sin\alpha}{L_z \cdot \cos\alpha} \\ m7 = 0 \\ m8 = 0 \\ m9 = \frac{1}{L_z} \\ m10 = m1 \cdot h_{ox} + m2 \cdot h_{oy} + m3 \cdot h_{oz} \\ m11 = m4 \cdot h_{ox} + m5 \cdot h_{oy} + m6 \cdot h_{oz} \\ m12 = m7 \cdot h_{ox} + m8 \cdot h_{oy} + m9 \cdot h_{oz} \end{cases} \tag{7}$$

The total magnetic field is invariable when the three-axis magnetic sensor changes its attitude in the uniform field. Therefore, the objective function can be expressed as

$$\hat{y} = \min \sum_{i=1}^{N} (|H_{\text{predicte}}|^2 - |H_{\text{theory}}|^2)^2 \tag{8}$$

where H_{predicte} refers to the predicted magnetic field with 9 unknowns, and H_{theory} represents the ideal magnetic field. The 9 unknowns are calculated by using the DA–LM algorithm. After that, the parameters of the nonorthogonal angle, scale factor, and zero bias are solved according to Equation (7). Finally, the error compensation process is completed.

3. Correction Algorithm

3.1. DA Algorithm

The DA algorithm is a bionics algorithm proposed in 2016 by Mirjalili, an Australian scholar [16]. The main idea comes from the static foraging behavior and dynamic migration behavior of dragonflies in nature. In a static group, to search for other flying prey, the dragonflies made up of small parts fly back and forth over a small area. The local motion during flight and the temporary mutation in the flight path are the characteristics of static groups. In dynamic groups, for a better living environment, large groups of dragonflies fly long distances to migrate or fly in a common direction [19]. According to dragonfly behavior, the algorithm can be divided into: separation, queuing, alliance, hunting for prey, and avoiding natural enemies. The separation weight, alignment weight, cohesion weight, prey weight factor, and natural enemy weight factor in the algorithm are used as the behavior degrees of dragonflies to update their position. The algorithm can improve the initial random parameters of a given problem and make it converge to the global optimum [20].

The mathematical expression of the DA algorithm is expressed as follows:

$$\Delta X_{t+1} = (s \cdot S_i + a \cdot A_i + c \cdot C_i + f \cdot F_i + e \cdot E_i) + \omega' \cdot \Delta X_t \tag{9}$$

where s refers to the separation weight, a represents the alignment weight, c denotes cohesion weight, f stands for the prey weight factor, e is the natural enemy weight factor, t refers to the current iteration number, ω' represents the inertia weight, S_i refers to the position vector of the separated behavior between the ith dragonfly, A_i represents the position vector of the queuing behavior of the ith dragonfly, C_i denotes the position vector of the alignment behavior of the ith dragonfly, F_i represents the position vector of the hunting behavior of the ith dragonfly, and E_i is the position vector of the avoidance behavior of the ith dragonfly.

In nature, dragonflies are in motion most of the time for survival. Therefore, their positions have to be updated in real time. Therefore, we obtain:

$$X_{t+1} = X_t + \Delta X_{t+1} \tag{10}$$

3.2. LM Algorithm

The LM is a modification of the Newton algorithm, which can be used to solve the problems that the Newton algorithm cannot guarantee, namely, that the search direction is always downward, and the Hessian matrix is always positively definite. The main idea of the Newton method is to use the first and second derivatives of iteration at Point x_k to make a quadratic approximation to the objective function. Then, using the minimum point as the new iterative point, the process is repeated until the approximate minimum point satisfying the requirements of accuracy is obtained. When the objective function $f : R^n \to R$ is second-order continuously differentiable, during the Taylor expansion of the function f at point x_k, the terms are ignored more than three times, and the quadratic approximation function can be obtained as follows:

$$f(x) \approx f(x_k) + (x - x_k)g(x_k) + \frac{1}{2}H(x_k)(x - x_k)^2 \tag{11}$$

where $g(x_k) = \nabla f(x_k)$ represents the first derivative of function f at point x_k, and $H(x_k) = \nabla^2 f(x_k)$ stands for the second derivative of function f at point x_k. If the first-order necessary condition of the local minimum is applied here, then we obtain:

$$\nabla f(x) = g(x_k) + H(x_k)(x - x_k) = 0 \tag{12}$$

If $H(x_k) > 0$, the minimum of function of f is

$$x_{k+1} = x_k - H(x_k)^{-1}g(x_k) \qquad (13)$$

which is the iterative formula of the Newton method. In case of univariate, if the second derivative of the function $f'' < 0$, the Newton method cannot converge to a minimum. While in case of multivariate, if the Hessian matrix $H(x_k)$ of the objective function is not positively definite, the search direction determined by using the Newton method is not necessarily going to be the direction in which the value of the objective function decreases. To solve this problem, the damping coefficient $\mu \geq 0$ is introduced, then the revised iteration formula is expressed as follows:

$$x_{k+1} = x_k - (H(x_k) + \mu_k I)^{-1}g(x_k) \qquad (14)$$

As long as μ is large enough, the search direction $d_k = -(H(x_k) + \mu_k I)^{-1}g(x_k)$ is ensured to be descending. Equation (14) is the iterative equation of the LM algorithm.

The goals and steps of the Algorithm 1: LM algorithm are as follows:

Algorithm 1. LM algorithm.

Goals : based on the function relationship $x = f(p)$, given the function f and noisy observation vector x, p can be estimated.

Step 1 : select the initial point p_0 and termination control constant ε and calculate $\varepsilon_0 = ||x - f(p_0)||$, $k := 0$, $\mu_0 = 10^{-3}$, and $v = 10$.

Step 2 : calculate the Jacobi matrix and $\overline{N_k} = \overline{J_k}J_k + \mu_k I$ and construct incremental normal equations $\overline{N_k} \bullet d_k = J_k^T \varepsilon_k$.

Step 3 : obtain the d_k from the incremental normal equations.

    (1) If $||x - f(p_k + d_k)|| < \varepsilon_k$, then let $p_{k+1} = p_k + d_k$ if $||d_k|| < \varepsilon$, stop iteration, and output the result. Otherwise, go back to Step 2.

    (2) If $||x - f(p_k + d_k)|| \geq \varepsilon_k$, then let $\mu_{k+1} = \mu_k v$, obtain the d_k again, and go back to Step 3 (1).

3.3. The Combination of DA and LM Algorithm

In solving the error parameters, we found that the number of least squares solutions were limited, while the LM algorithm can be used to solve multiple parameters simultaneously. However, the disadvantage of the LM algorithm is that it may fall into the local minima if the initial value is not appropriately selected, which has a serious impact on the accuracy of problem solving. To overcome this drawback, we used a global optimization algorithm to find the suitable initial value for the LM algorithm. As a global optimization algorithm, the DA algorithmhas the advantages of its strong global optimization ability. However, it has low accuracy in local optimization. To tackle this problem, this paper proposes to combine the DA algorithm and the LM algorithm. Firstly, this paper usedthe DA algorithm's global optimization ability to find the global minimum point. Then, it took the output parameter value of the DA algorithm as the initial value of the LM algorithm for local optimization. The Figure 2 shows the flow chart of DA–LM algorithm.

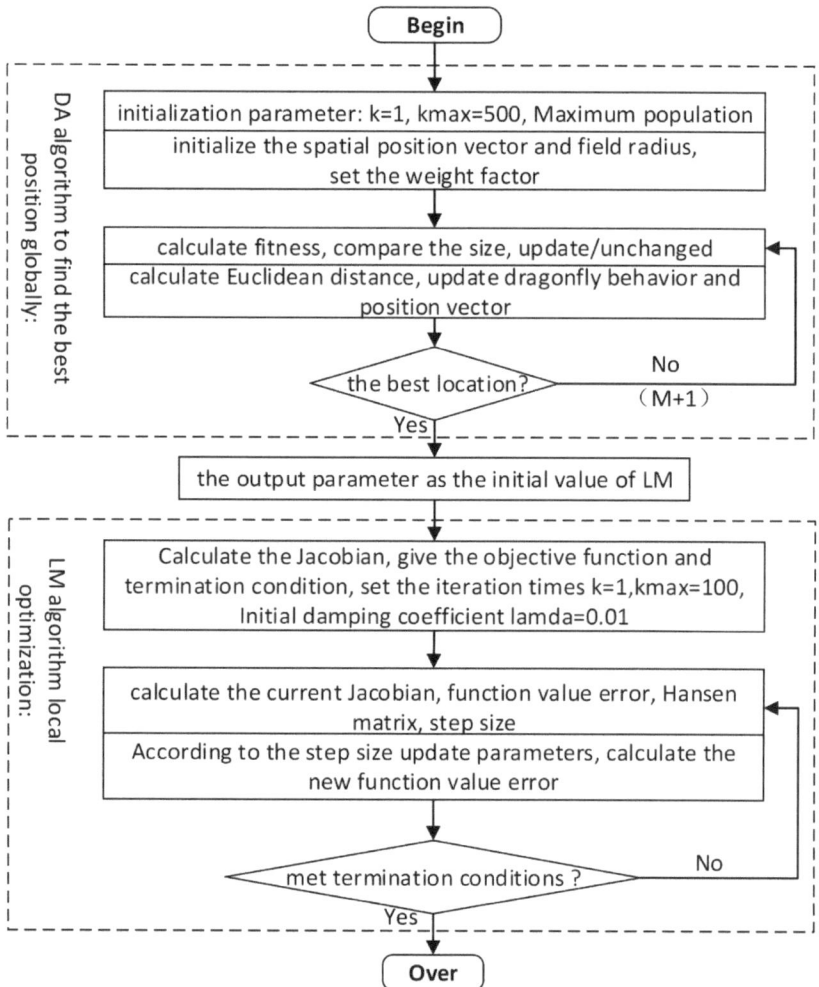

Figure 2. The flow chart of DA–LM algorithm.

4. Simulated Analysis

Matlab is used for data simulation to verify the correctness and effectiveness of the proposed method based on the DA–LM algorithm in this paper. In the simulation analysis, the total magnetic field was set as 50,000 nT. Using the random function of Matlab to generate 96 groups of different angles, the rotation of the sensor was simulated, and then, the 96 groups of the ideal three-axis sensor rotation data H were obtained. The measured value was simulated by using Equation (4). The error parameter of the three-axis sensor is shown by the preset value in Table 1. The H1 three components of the magnetic field and the H1 total magnetic field are shown in Figure 3. The abscissa axis indicates the sampling points. The ordinate axes of (a), (b), and (c) indicate the components of the measured values H1 in x, y, and z, and the vertical axis of (d) represents the total magnetic field of the measurement H1. In addition, the red line refers to the ideal total magnetic field.

Table 1. Comparison of preset and estimate parameters.

Error Term		Preset Value	LM Estimated Value	DA–LM Estimated Value
Nonorthogonal /(°)	α	0.000622	0.031693	0.000622
	β	0.000332	−0.014156	0.000332
	γ	−0.000076	0.035489	−0.000076
Scale Factor	L_x	1.002685	1.002685	1.002685
	L_y	1.002853	1.002853	1.002853
	L_z	1.002964	1.002964	1.002964
Zero Offset /(nT)	h_x	−23.210025	−23.210029	−23.2100249999997
	h_y	−44.730353	−44.744776	−44.7303529999983
	h_z	−170.944506	−170.962902	−170.9445060000002

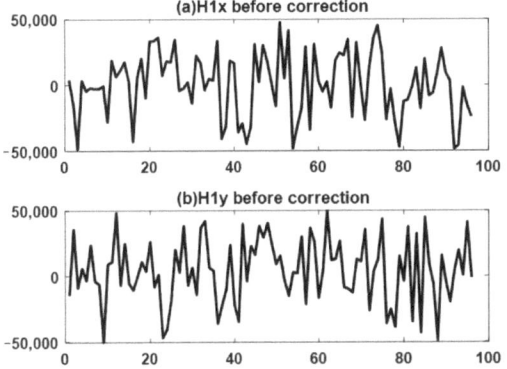
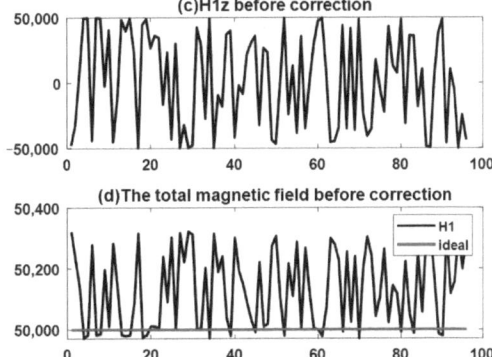

Figure 3. The components and total magnetic field of H1 before correction. (**a**) The component of the measured values of x-axis; (**b**) The component of the measured values of y-axis; (**c**) The component of the measured values of z-axis; (**d**) The total magnetic field of measured values and ideal values.

By substituting, the ideal magnetic field and H1 into the objective function (Equation (8)) and using the LM and DA–LM respectively, the comprehensive parameters P and Q were solved. According to the relations among P, Q and the error parameters of the three-axis magnetic sensor, the error parameters were inversely solved. Table 1 shows the comparison between the error parameters obtained by inverse solution and the preset values. Finally, the magnetic field was corrected. The errors of the components and total magnetic field before and after correction are shown in Figure 4 where the abscissa axis represents the sampling point, and the ordinate axes of (a), (b), (c), and (d) stand for the errors of the x-axis, y-axis, z-axis, and the total magnetic field, respectively. The black lines show the errors before correction. And the blue lines represent the errors after correction.

Seen from Figure 4, the errors of the x-axis, y-axis, z-axis, and the total magnetic field between the measured and ideal values are 200 nT, 200 nT, 350 nT, and 350 nT, respectively. After the DA–LM algorithm correction, all the errors were reduced to 0 nT. The parameter estimates are shown in Table 1.

It can be seen from the comparison among the preset parameter values, LM estimate values, and DA–LM estimate values in Table 1 that the LM algorithm alone cannot accurately estimate the error parameters, especially when the nonorthogonal angle is very small. The reason is that the initial value is given empirically when the LM algorithm is used alone. If the initial value is given improperly, the minima will be trapped in the local minimum. The error parameters calculated by using the DA–LM algorithm were almost the same as the preset value, with the error accuracy of estimation of 10^{-6} magnitude. The simulation results show that the global optimization ability of the DA algorithm can help the LM algorithm find the most suitable initial value and effectively prevent the LM algorithm from falling into the local minimum.

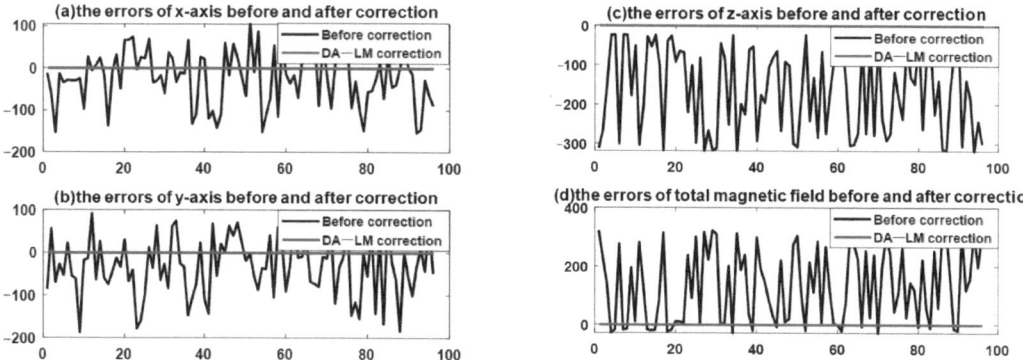

Figure 4. The errors of components and total magnetic field before and after correction. (**a**) The errors of the *x*-axis; (**b**) The errors of the *y*-axis; (**c**) The errors of the *z*-axis; (**d**) The errors of measured total magnetic field before and after correction.

5. Experimental Verification

In addition to its own error, the sensor is also susceptible to many factors, for instance, magnetic diurnal variation. To mitigate the influence of the magnetic diurnal variation on the magnetic field, experiments were carried out after 12 p.m. The ideal total magnetic field H was measured by using an optical pump magnetometer with high precision, as shown in Figure 5. The actual measurement value H1 was obtained by using the three-axis magnetometer MAG648. The experiment platform is as shown in Figure 6. In Figure 7, the abscissa axis represents the sampling point of the measured data, and the ordinate axis stands for the total value of the magnetic field. The red line refers to the ideal total magnetic field modulus, while the black line shows the measurement total magnetic field H1 before correction. The blue line refers to the total magnetic field after LM correction, and the green line denotes the total magnetic field after DA–LM correction. It is obvious that the DA–LM correction effect outperforms that of LM correction.

Figure 5. The optical pump magnetometer with high precision.

Figure 6. The experiment platform with the three-axis magnetometer MAG648.

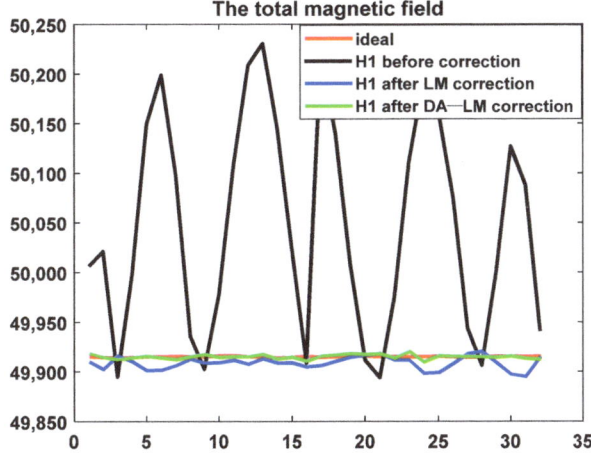

Figure 7. The total magnetic field before and after correction.

In Figure 8, the abscissa axis refers to the sampling point, and the ordinate axis represents the error between the total value measured data and the ideal total value. In addition, the value is reduced from 300 nT to 5 nT and 20 nT after DA–LM and LM correction, respectively. We drew a conclusion that the steering error was significantly suppressed. The parameters shown in Table 2 were solved by using the DA–LM method and LM method, respectively. Table 3 shows the comparison between the mean value and the root mean square error. It can be seen from the Table that the mean value and root mean square error were reduced after DA–LM correction.

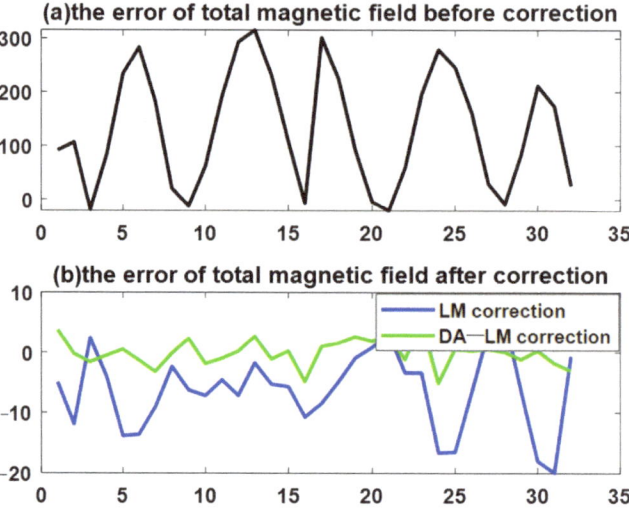

Figure 8. The error of total magnetic field before and after correction. (**a**) The error of total magnetic field before correction; (**b**) The error of total magnetic field after correction, and the blue line represents the LM correction, the green line represents the DA–LM correction.

Table 2. The parameters obtained by calculation.

Error Term		LM Estimated Value	DA–LM Estimated Value
Nonorthogonal	α	0.002632	−0.000478
	β	−0.004717	−0.000390
	γ	0.019320	−0.000028
Scale Factor	L_x	1.002880	1.002692
	L_y	1.003050	1.002862
	L_z	1.002985	1.002974
Zero Offset	h_x	−23.365514	−23.382624
	h_y	−45.282038	−45.178011
	h_z	−171.458722	−171.802028

Table 3. Comparison of statistical characteristics before and after DA–LM correction.

Evaluation Item	RMS (nT/m)	RMSE (nT/m)
Before Correction	−131.807381	109.523401
After Correction	0.000247	2.285982

6. Conclusions

Geomagnetic error correction is the key to obtaining high-precision geomagnetic information and the premise of magnetic anomaly location. Error correction has always been the focus of research on magnetic measurement, and many correction algorithms have been proposed under different application backgrounds. This paper analyzed the error sources of magnetic sensors, proposed a method combing the DA algorithm and the LM algorithm to iteratively solve the parameters in the error model, and finally completed the error compensation. The following conclusions are drawn: (1) The method combining the DA and LM algorithms can accurately calculate the error parameters, which solves the problem that it may easily fall into the local minimum in the iterative process by using the LM algorithm alone; (2) The simulation results show that the proposed method can accurately estimate the error parameters and has a good correction effect on the components and total magnetic field, thereby proving the effectiveness of the proposed algorithm. It can be seen from the experimental results that this method can estimate the error parameters of sensors well, with the RMS value reduced from -131.807381 nT/m to 0.000247 nT/m and the RMSE value from 109.523401 nT/m to 2.285982 nT/m. The experimental data were only calibrated for the total magnetic field due to the unavailability of the ideal three-component values with high accuracy. In the event the ideal three-component can be obtained, this method can also be applied to three-component correction.

Author Contributions: Conceptualization, C.L. and L.Y.; methodology, C.L. and L.Y.; software, C.L., L.Y. and S.Z.; validation, C.L. and L.Y.; investigation, C.L., L.Y., S.Z., S.X., X.T. and Y.L.; data curation, H.C., S.Z. and C.X.; writing–original draft preparation, C.L.; writing–review and editing, C.L. and L.Y.; supervision, L.Y., C.X. and H.C. All authors have read and agreed to the published version of the manuscript.

Funding: This research received no external funding.

Institutional Review Board Statement: Not applicable.

Informed Consent Statement: Not applicable.

Data Availability Statement: All of the data in our paper are not publicly archived data.

Conflicts of Interest: The authors declare no conflict of interest.

References

1. Sun, X.J.; Kou, J.; Zhang, X.N.; Li, J. Research Progress in Geomagnetic Navigation. *Navig. Control* **2016**, *15*, 1–6.
2. Pan, Q.J.; Ma, W.M.; Zhao, Z.H.; Kang, J. Development and Application of Measurement Method for Magnetic Field. *Trans. China Electrotech. Soc.* **2005**, 7–13. [CrossRef]
3. Ma, J.Y. Application of high precision magnetic prospecting in regional investigation. *World Nonferr. Met.* **2018**, *33*, 148–149.
4. Bilukha, O.O.; Brennan, M.; Anderson, M. Injuries and Deaths From Landmines and Unexploded Ordnance in Afghanistan 2002–2006. *JAMA* **2007**, *298*, 516–518. [CrossRef] [PubMed]
5. Pei, Y.L.; Liu, B.H.; Zhang, G.E.; Liang, R.C.; Li, X.S. Application of Magnetic Method to Ocean Engineering. *Adv. Mar. Sci.* **2005**, 114–119.
6. Song, H.; Zou, L.; Zhang, X.Q.; Zhang, L.; Zhao, T. Inter-Turn Short-Circuit Detection of Dry-Type Air-Core Reactor Based on Spatial Magnetic Field Distribution. *Trans. China Electrotech. Soc.* **2019**, *34* (Suppl. 1), 105–117.
7. Pham, D.M.; Aziz, S.M. A real-time localization system for an endoscopic capsule using magnetic sensors. *Sensors* **2014**, *14*, 20910–20929. [CrossRef] [PubMed]
8. Shi, G.; Li, X.S.; Li, X.F.; Liu, Y.X.; Kang, R.Q.; Shu, X.Y. Equivalent two-step algorithm for the calibration of three-axis magnetic sensor in heading measurement system. *Chin. J. Sci. Instrum.* **2017**, *38*, 402–407.
9. Liu, J.J.; Chen, L.; Wang, D.J.; Qi, H.X. Correction method of 3-axis magnetic sensor. *Mod. Electron. Tech.* **2018**, *41*, 179–181+186.
10. Li, X.; Li, Z. Dot product invariance method for the calibration of three-axis magnetometer in attitude and heading reference system. *Chin. J. Sci. Instrum.* **2012**, *33*, 1813–1818.
11. Li, X.; Song, B.Q.; Wang, Y.J.; Niu, J.H.; Li, Z. Calibration and Alignment of Tri-Axial Magnetometers for Attitude Determination. *IEEE Sens. J.* **2018**, 7399–7406. [CrossRef]
12. Luo, J.G.; Li, H.B.; Liu, J.X.; Li, H.H.; Zhang, F. Study on Error Correction Method of Three-axis Fluxgate Magnetometer. *Navig. Control* **2019**, *18*, 52–58.
13. Pang, H.F.; Luo, S.T.; Chen, D.X.; Pan, M.C.; Zhang, Q. Error Calibration of Fluxgate Magnetometers in Arbitrary Attitude Situation. *J. Test Meas. Technol.* **2011**, *25*, 371–375.
14. Wang, Q.B.; Zhang, X.M. Research of fast calibration method of three-axis magnetic sensor. *China Meas. Test* **2017**, *43*, 35–39.
15. Pang, X.L.; Lin, C.S. Calibration Coefficients Solving of Three-Axis Magnetic Sensor Based on Genetic Algorithm. *J. Detect. Control* **2017**, *39*, 42–45+51.
16. Mirjalili, S. Dragonfly algorithm: A new meta-heuristic optimization technique for solving single-objective, discrete, and multi-objective problems. *Neural. Comput. Appl.* **2016**, *27*, 1053–1073. [CrossRef]
17. Zhang, H.Y.; Geng, Z. Novel interpretation for Levenberg-Marquardt algorithm. *Comput. Eng. Appl.* **2009**, *45*, 5–8.
18. Zhang, J.; Chen, J.; Yang, S.H. The Study of Calibration Method of Magnetic Resistance Sensor. *J. Proj. Rocket. Missiles Guid.* **2010**, *30*, 46–48.
19. Wu, W.M.; Wu, W.Y.; Lin, Z.Y.; Li, Z.X.; Fang, D.Y. Dragonfly algorithm based on enhancing exchange of individuals' information. *Comput. Eng. Appl.* **2017**, *53*, 10–14.
20. Qiao, M.Y.; Xu, C.K.; Tang, X.X.; Gao, Y.F.; Shi, J.K. Application Research of DA-LM Algorithm in Error Correction of MEMS Accelerometer. *Chin. J. Sens. Actuators* **2021**, *34*, 223–231.

MDPI AG
Grosspeteranlage 5
4052 Basel
Switzerland
Tel.: +41 61 683 77 34

Metals Editorial Office
E-mail: metals@mdpi.com
www.mdpi.com/journal/metals

Disclaimer/Publisher's Note: The title and front matter of this reprint are at the discretion of the Guest Editor. The publisher is not responsible for their content or any associated concerns. The statements, opinions and data contained in all individual articles are solely those of the individual Editor and contributors and not of MDPI. MDPI disclaims responsibility for any injury to people or property resulting from any ideas, methods, instructions or products referred to in the content.

www.ingramcontent.com/pod-product-compliance
Lightning Source LLC
LaVergne TN
LVHW072358090526
838202LV00019B/2571